Getting Started with Stata
for Windows®

Stata Press
College Station, Texas

Stata Press, 702 University Drive East, College Station, Texas 77840

The suggested citation for this software is

StataCorp. 1999. *Stata Statistical Software: Release 6.0*. College Station, TX: Stata Corporation.

Contents

About this manual

Stata for Windows® is discussed in this manual. Stata for Macintosh users should see *Getting Started with Stata for Macintosh*; Stata for Unix users, see *Getting Started with Stata for Unix*.

This manual is intended both for people completely new to Stata and for experienced Stata users new to Stata for Windows. Previous Stata users will also find it helpful as a tutorial on some new features in Stata for Windows.

Each numbered chapter opens with a short summary of the contents of that chapter. New users can get a 15-minute course on Stata by reading all the summary pages first, and then go back and read all or part of the text more thoroughly. Previous users will find the summaries a useful reference for Stata's basic commands.

Following the numbered chapters are five appendices with information specific to Stata for Windows.

Throughout this manual, you will see things like "see [R] **ci**". This is a reference to the entry for **ci** in the 4-volume *Stata Reference Manual*. You will also see things like "see [G] **printing Windows**", which is a reference to the *Stata Graphics Manual*, and "see [U] **12 The Break key**", which is a reference to the *Stata User's Guide*. Finally, you may see things like "[GSW] **D. Setting the size of memory**", which is a reference to Appendix D at the end of this manual.

We provide technical support to registered Stata users. Chapter 3 of this manual describes the sources of information available to help you learn about Stata's commands and features. One of these sources is the Stata web site (http://www.stata.com). Half of the web site is dedicated to user support. You will find answers to frequently asked questions (FAQs) as well as much other useful information for users. If, after looking at the Stata web site and the other sources of information described in Chapter 3, you still have questions, you can contact us as described in [U] **2.8 Technical support** in the *Stata User's Guide*.

1 Installation

Stata for Windows

Description

Version of Stata for running under Windows.
Stata for Windows is available for Windows 98/95/NT or Windows 3.1.

(Versions for Macintosh and Unix are also available.)

Intercooled Stata

Professional version of Stata.
Very fast.
Number of observations limited only by computer memory.
Runs on 80386 or better.
Requires math coprocessor, either on-chip (Pentium II; Pentium; 80486DX) or added (80486SX with 80487; 80386 with 80387).

Pseudo-Intercooled
(no math coprocessor)

Same as Intercooled Stata, but slower.
Does not require math coprocessor.
Runs on 80386 or better (80486SX; 80386 without 80387).

Note: We recommend you add a math coprocessor and run Intercooled Stata.

Small Stata

Stata for small computers.
Datasets restricted to approximately 300K bytes.
Maximum of 99 variables and approximately 1,000 observations.
Does not require math coprocessor.
Slower than Intercooled Stata.
Runs on 80386 or better.

You received

Whether you purchased Intercooled or Small Stata:

Single sheet *License and Authorization Key*.

Disks labeled either *Stata for Windows 98/95/NT* or *Stata for Windows 3.1*, depending on which you purchased.

Getting Started with Stata for Windows (this book).

Stata User's Guide.

4-volume *Stata Reference Manual*.

Stata Graphics Manual.

Registration Card.

Stata for Windows, continued

What to install   Look at your *License and Authorization Key*.

**If you have an Intercooled Stata license
and a math coprocessor:**

Install Intercooled Stata.
May install a second copy—Intercooled Stata
or Small Stata—on a second computer (e.g., office
and home; but no simultaneous use).

**If you have an Intercooled Stata license
and no math coprocessor (some Windows 3.1 users):**

Install Intercooled Stata.
Pseudo-Intercooled (no math coprocessor) Stata
will also be automatically installed. Run
Pseudo-Intercooled Stata rather than Intercooled.
May install a second copy—Intercooled Stata
or Small Stata—on a second computer (e.g., office
and home; but no simultaneous use).

If you have a Small Stata license:

Install Small Stata.
Do not install Intercooled Stata; your license
will not let you run it.
May install Small Stata on a second computer
(e.g., office and home; but no simultaneous use).

Important   Do not lose your paper license. Keep it in a safe place.
You may need it again in the future.

Before you install

Before you begin the installation procedure:

1. Make sure you have the Stata for Windows disks.

    a. If you use Windows 98, 95, or NT, make sure you have the Stata for Windows 98/95/NT disks.

    b. If you use Windows 3.1 or 3.11, make sure you have the Stata for Windows 3.1 disks.

2. Make sure you have a License and Authorization Key.

3. Decide whether you are installing Intercooled or Small Stata (see previous page).

4. Decide where you want to install the Stata software. We recommend `C:\STATA`.

5. Decide where you want to set the working directory. This should be different from the installation directory so that files that you create will not get mixed up with Stata's files. We recommend `C:\DATA`.

6. If you already have an old version of Stata on your system, decide whether you want to keep it or delete it. If you want to install the new version in the same directory, you must let the installation program remove the old version. If you want to keep the old version, you must install the new version in a different directory.

7. You are ready to install; turn to the appropriate section:

    Installation and upgrades for Windows 98/95/NT on page 4

        or

    Installation and upgrades for Windows 3.1 on page 7.

Note

The installation procedure does not change your `AUTOEXEC.BAT` or `CONFIG.SYS` files.

Upgrade or update?

If you use Stata 5.0 or an earlier release, and you wish to upgrade to Stata 6.0, this is the chapter for you. If you have already installed Stata 6.0, and you wish to install the latest updates to Stata 6.0, see Chapter 20.

Installation and upgrades for Stata for Windows 98/95/NT

Please be sure that Windows is installed and properly running before attempting to install Stata for Windows 98/95/NT. Have your Stata License and Authorization Key with you.

1. Insert *Stata for Windows 98/95/NT* Disk 1 in the floppy disk drive.

2. Click on the **Start** button, select **Settings**, and choose **Control Panel**.

3. Double-click the **Add/Remove Programs** icon.

4. In the **Add/Remove Programs** dialog box, press the **Install...** button.

5. A dialog will appear asking you to insert the first installation floppy. Assuming that you have done so, press the **Next** button.

6. The Stata for Windows 98/95/NT installation will begin. You will see a couple of opening screens of information. After you read these screens carefully, press the **Next** button.

7. The installation program will ask you where you want to install Stata.

   a. We recommend that you choose the default directory—`C:\STATA`.

   b. When you have chosen an installation directory, press the **Next** button.

   c. If an older Stata is already installed in this directory, you will later be asked if you want to uninstall the older version. Only the official distribution files from the older version of Stata will be deleted in case you have some personal files in that directory. If you do not want to delete the previous version, you should choose a different directory in which to install Stata now.

Installation and upgrades for Stata for Windows 98/95/NT, continued

8. At the **Select Components** dialog, you can choose which flavor of Stata to install.

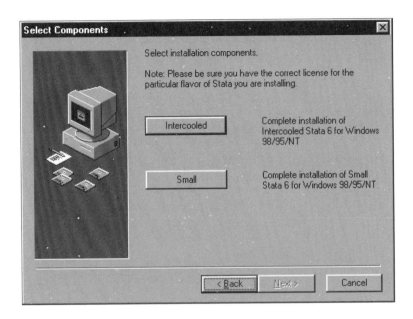

    a. Look at your License and Authorization Key. If it is a Small Stata license, press the **Small** button. Intercooled Stata will not run under a Small Stata license.

    b. If you have an Intercooled Stata license, press the **Intercooled** button.

9. The installation program will prompt you to press **Next** to begin the installation.

    a. When you press **Next**, the installation program will determine if there is a previous version of Stata installed in the directory that you selected in step 7.

    b. If there is a previous version, the installation program will ask you if it is okay to uninstall the previous version. Only the official distribution files from the older version of Stata will be deleted in case you have some personal files in that directory. If you do not want to delete the previous version, you should exit the installation, start over, and choose a different installation directory in step 7.

10. The installation program will then ask you to select the default working directory.

    a. The default working directory is where your datasets, graphs, and other Stata-related files will be stored.

    b. We recommend that you choose the default directory—`C:\DATA`.

    c. When you have chosen a default working directory, press the **OK** button.

11. The installation program will copy Stata to your hard disk. You may be prompted to insert additional Stata disks at this time. Do not worry if you are not prompted for all Stata disks; some files may not be needed depending on whether you are installing Intercooled or Small Stata.

Installation and upgrades for Stata for Windows 98/95/NT, continued

12. When the installation is complete, it will prompt you to press the **Finish** button to exit the installation.

13. The first time you run Stata for Windows 98/95/NT, it will prompt you for the information on your License and Authorization Key. Press *Tab* to move from one line to the next; press *Enter* or the **OK** button when finished. You must enter something for all fields in the dialog before you can continue. If you make a mistake typing the codes, you will be prompted to try again.

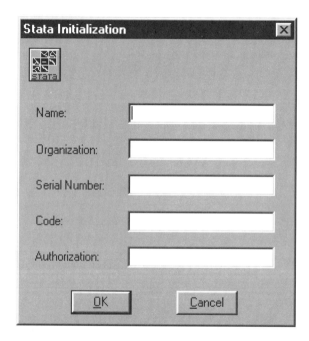

14. The installation is now complete. Turn to Chapter 2 for information on starting and stopping Stata for Windows 98/95/NT.

Installation and upgrades for Stata for Windows 3.1

Please be sure that Windows 3.1 is installed and properly running before attempting to install Stata for Windows 3.1. Have your Stata License and Authorization Key with you.

1. Insert *Stata for Windows 3.1* Disk 1 in the floppy disk drive.

2. From the Windows Program Manager, select **Run...** from the **File** menu.

3. At the **Command Line** prompt, enter `a:setup` (or `b:setup` if you use the B drive).

4. A dialog will appear asking you to insert the first installation floppy. Assuming that you have done so, you should press the **Next** button.

5. The Stata for Windows 98/95/NT installation will begin. You will see a couple of opening screens of information. After you read these screens carefully, press the **Next** button.

6. The installation program will ask you where you want to install Stata.

   a. We recommend that you choose the default directory—`C:\STATA`.

   b. When you have chosen an installation directory, press the **Next** button.

   c. If an older Stata is already installed in this directory, you will later be asked if you want to uninstall the older version. Only the official distribution files from the older version of Stata will be deleted in case you have some personal files in that directory. If you do not want to delete the previous version, you should choose a different directory in which to install Stata now.

Installation and upgrades for Stata for Windows 3.1, continued

7. At the **Select Components** dialog, you can choose which flavor of Stata to install.

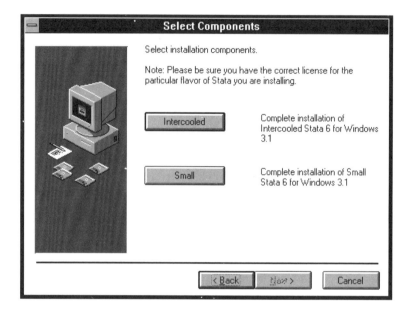

a. Look at your License and Authorization Key. If it is a Small Stata license, press the **Small** button. Intercooled Stata will not run under a Small Stata license.

b. If you have an Intercooled Stata license, press the **Intercooled** button. This will install both Intercooled and Pseudo-Intercooled Stata. If your computer has a math coprocessor, run Intercooled Stata. If your computer does not have a math coprocessor, run Pseudo-Intercooled Stata.

8. The installation program will prompt you to press **Next** to begin the installation.

a. When you press **Next**, the installation program will determine if there is a previous version of Stata installed in the directory you selected in step 6.

b. If there is a previous version, the installation program will ask you if it is okay to uninstall the previous version. Only the official distribution files from the older version of Stata will be deleted in case you have some personal files in that directory. If you do not want to delete the previous version, you should exit the installation, start over, and choose a different installation directory in step 6.

9. The installation program will then ask you to select the default working directory.

a. The default working directory is where your datasets, graphs, and other Stata-related files will be stored.

b. We recommend that you choose the default directory—C:\DATA.

c. When you have chosen a default working directory, press the **OK** button.

Installation and upgrades for Stata for Windows 3.1, continued

10. The installation program will copy Stata to your hard disk. You may be prompted to insert additional Stata disks at this time. Do not worry if you are not prompted for all Stata disks; some files may not be needed depending on whether you are installing Intercooled or Small Stata.

11. When the installation is complete, it will prompt you to press the **Finish** button to exit the installation.

12. The first time you run Stata for Windows 3.1, it will prompt you for the information on your License and Authorization Key. Press *Tab* to move from one line to the next; press *Enter* or the **OK** button when finished. You must enter something for all fields in the dialog before you can continue. If you make a mistake typing the codes, you will be prompted to try again.

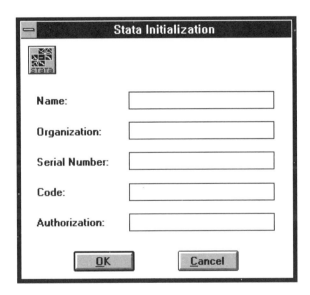

13. The installation is now complete. Turn to Chapter 2 for information on starting and stopping Stata for Windows 3.1.

Notes

2 Starting and stopping Stata

Testing the installation

Below we assume Stata is installed. When you launch Stata from the Windows 98/95/NT **Start** menu, or from the Windows 3.1 Program Manager by double-clicking on the Stata icon, the first screen you will see is

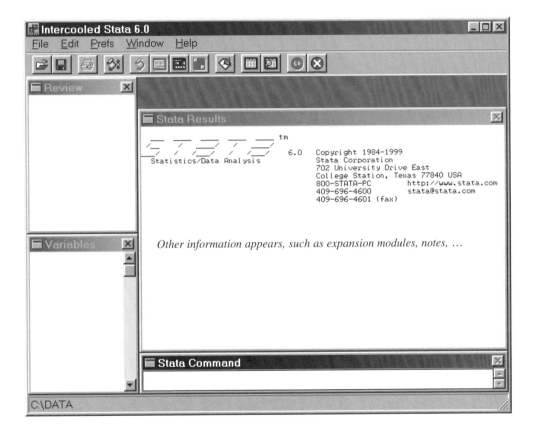

If you see a message box telling you that Stata could not find the license file or you receive any other error when you try to start Stata, please see [GSW] **C. Troubleshooting starting and stopping Stata** at the back of this book.

Note

There are other ways to start Stata than from the **Start** menu in Windows 98/95/NT or from the Program Manager in Windows 3.1. You may find one of these ways more convenient. Please read [GSW] **A. Starting and stopping Stata for Windows 98/95/NT** or [GSW] **B. Starting and stopping Stata for Windows 3.1** at the end of this manual for more information.

The Stata windows

Note

The screenshot below is of Stata for Windows 98/95/NT. Stata for Windows 3.1 looks slightly different. It has words on its buttons rather than icons; see page 181 for a screenshot of Stata for Windows 3.1.

Past commands appear here   **Results are displayed here**

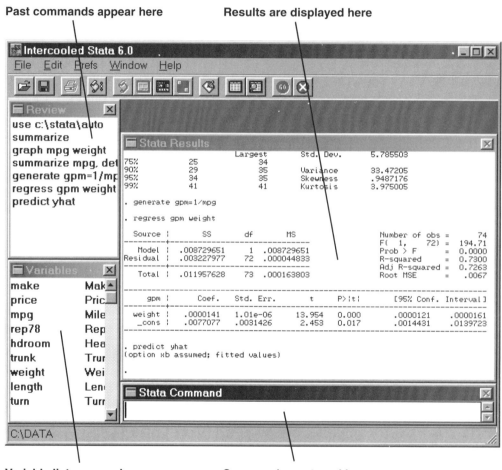

Variable list appears here   **Commands are typed here**

The Stata toolbar

Note

> Stata for Windows 3.1 does not have all the buttons below. In addition, Stata for Windows 3.1 has words on its buttons rather than icons.

Stata for Windows 98/95/NT has thirteen buttons. If you ever forget what a button does, hold the mouse pointer over a button for a moment and a box will appear with a description of that button.

Open

> Open a Stata dataset.

Save

> Save to disk the Stata dataset currently in memory.

Print Graph/Print Log

> Print a graph or log.

Log Start/Stop/Suspend (**Log...** in Windows 3.1)

> Start a new log, append to an existing log, and stop or suspend the current log. See Chapter 16 for an explanation of log files.

Bring Log to Front

> Bring the Log window to the front of the other Stata windows.

Bring Dialog to Front (**Dialog** in Windows 3.1)

> Bring the Dialog window to the front of the other Stata windows. See [R] **window dialog** for more information.

Continued

The Stata toolbar, continued

Bring Results to Front (**Results** in Windows 3.1)
Bring the Results window to the front of the other Stata windows.

Bring Graph to Front (**Graph** in Windows 3.1)
Bring the Graph window to the front of the other Stata windows. See Chapter 15 for more information about graphs.

Do-file Editor
Open the do-file editor or bring the Do-file Editor window to the front of the other Stata windows. See Chapter 14 for more information.

Data Editor (**Editor** in Windows 3.1)
Open the data editor or bring the Data Editor window to the front of the other Stata windows. See Chapter 4 for more information.

Data Browser (**Browse** in Windows 3.1)
Open the data browser or bring the Data Browser window to the front of the other Stata windows. See Chapter 8 for more information.

Clear –more– Condition (**More** in Windows 3.1)
Tell Stata to continue when it has paused in the middle of long output. See Chapter 10 for more information.

Break (**Break** in Windows 3.1)
Stop the current task in Stata. See Chapter 10 for more information.

Verifying the installation

Since you have just installed Stata for Windows, you should verify that the installation was performed correctly. Stata includes a command called verinst. Type verinst in the Command window and press *Enter*.

You should see output in the Stata Results window informing you that Stata is correctly installed.

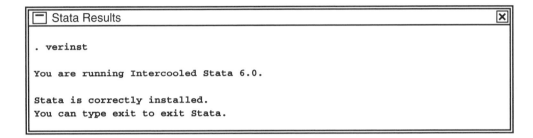

If you see the error message below complaining that the command was not found,

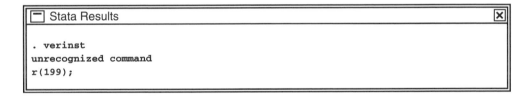

then not all of Stata is installed. Go back to the installation chapter and reinstall Stata.

If Stata informed you that it was correctly installed, you can do one of the following to exit:

1. Type exit in the Command window and press *Enter*. In the future, you may need to type exit, clear.

2. Choose **Exit** from the **File** menu.

3. Click on the close box (the box with an **X** in the upper right-hand corner of the Stata window). If you are using Windows 3.1, double-click on the control-menu box (the box in the upper left-hand corner of the Stata window).

The working directory

If you look at the screenshot on page 11, you will notice a status window at the bottom of the screen that contains C:\DATA. Stata is telling you that C:\DATA is the current working directory. (In Stata for Windows 3.1, the working directory is displayed in the title bar of the Command window.)

The working directory is where graphs and datasets will be saved unless you specify another directory. If you use Stata for Windows 98/95/NT, see the next page for instructions on changing the working directory. If you use Stata for Windows 3.1, see page 18.

Once you have started Stata, you can change the working directory with the cd command. See [R] **cd** in the *Reference Manual* for full details.

Stata always displays the name of the working directory, so it is easy to tell where your graphs and datasets will be saved.

Increasing the amount of memory Intercooled Stata uses (98/95/NT)

You can change the amount of memory Stata uses once it is running; see [GSW] **D. Setting the size of memory**. If you do that, the change is effective only during that session. Or you can change the amount of memory used every time you invoke Stata.

To reset the amount of memory Intercooled or Pseudo-Intercooled Stata uses (Small Stata cannot be changed):

1. Navigate to the location of the **Start** menu's Stata shortcut on your hard disk.

   a. Select **Run...** from the **Start** menu.

   b. Windows 98 and 95 users type `"C:\WINDOWS\Start Menu\Programs\Stata"`

   c. Windows NT users type `"C:\WINNT\Profiles\All Users\Start Menu\Programs\Stata"`

   d. The double-quotes are important; do not omit them.

   e. Press *Enter*.

2. You will find one, two, or three icons in this folder depending on how many flavors of Stata you installed. Click once on the one you wish to change.

3. Pull down **File** and choose **Properties**.

4. Click on the **Shortcut** tab. The **Target** edit field is the Stata command line. It probably says something like

   `C:\stata\WSTATA.EXE /k1000`

   The `/k1000` means Stata is to allocate 1,000K (or 1 megabyte) to its data area. If you were to change the option to `/k2000`, Stata would allocate 2 megabytes. If you were to change it to `/k64000`, Stata would allocate 64 megabytes. Change the number as you wish. The number need not be a multiple of 1,000.

5. The **Start in** edit field says where Stata is to start. It probably contains

   `C:\DATA\`

   This sets Stata's initial working directory. Change it if you wish.

6. Click on **OK** and you are done. The next time you start Stata from the **Start** button, it will start with the size and in the directory you specified.

Note

See [GSW] **A.8 Making shortcuts** and [GSW] **A.6 Specifying the amount of memory allocated** for more information on creating additional shortcuts to Stata with different amounts of memory. Also, see [GSW] **D. Setting the size of memory** for information on changing the amount of memory allocated to Stata while Stata is running.

Increasing the amount of memory Intercooled Stata uses (Windows 3.1)

You can change the amount of memory used every time you invoke Stata.

To reset the amount of memory Intercooled or Pseudo-Intercooled Stata uses (Small Stata cannot be changed):

1. In the Program Manager, click once on the Stata icon.

2. Choose **Properties** from the **File** menu.

   You will see four different lines in the properties that look something like

Description:	Intercooled Stata
Command Line:	C:\stata\wstata /k1000
Working Directory:	C:\data
Shortcut Key:	None

   It is the /k# on the **Command Line** that determines the amount of memory Intercooled Stata uses. The amount of memory is in units of kilobytes (1,000 kilobytes is 1 megabyte). So, you could change Intercooled Stata to come up with 4 megabytes by changing that line to look like

Command Line:	C:\stata\wstata /k4000

3. Some users will want to have Stata come up with different amounts of memory under different circumstances. This is easily done using multiple icons. See [GSW] **B.8 Creating multiple Stata icons** and [GSW] **B.6 Specifying the amount of memory allocated** for details.

4. Note that, if you want, you can also change the working directory by editing the **Working Directory** line.

3 Help

On-line help

In this chapter, you will learn:

Stata for Windows has a comprehensive **Help** system.

You can have a Help window open while you enter commands
in the Command window.

When you choose **Help** from the main menu bar,
you get a menu from which you can:
- See the help table of contents
- Search for help entries on a topic
- Get help for a Stata command
- List the latest features in Stata
- Install the latest official updates from diskette (or download them from the web
if you use Stata for Windows 98/95/NT)
- Install new user-written or *Stata Technical Bulletin* programs for Stata from
diskette (or download them from the web if you use Stata for Windows 98/95/NT)
- Go straight to important points on the Stata web site if you use Stata for
Windows 98/95/NT

Choosing **Search...** from the **Help** menu allows you to enter keywords
and produces a screen containing:
- Hypertext links (clickable green words) that will take you to the help files for the
appropriate Stata commands
- Plus references to the topic in the *Reference Manual* and *Graphics Manual*
- Plus references to the topic in the *User's Guide*
- Plus references to the topic in the *Stata Technical Bulletin*
- Plus FAQs on Stata's web site dealing with the topic

Example:
- Select **Search...** from the **Help** menu
- Enter `regression` and click **OK**

and you will get all the references to regression in the *Reference Manual* and the *User's
Guide* and a list of all the Stata commands that relate to regression:
- Stata commands like `areg`, `cnreg`, `cnsreg`, ... will appear in green

When you position the mouse pointer near a hypertext link, the pointer will change to
a hand. If you click while the hand is pointing at a command name like `areg`, you will
go to
- The help file for `areg`

Continued

Help, continued

Multiple topic words are allowed with **Search**
Adding topic words will narrow the search; for example:
- Enter `regression residuals`

Choosing **Contents** from the **Help** menu gives a listing of Stata's help table of contents.
- You may choose from the links on this page to view help for a particular command
- Or you may enter the full name of a Stata command in the edit field at the top of the Help window

Example:
- Click on the edit field at the top of the Help window
- Enter `ttest` (`ttest` is a Stata command) and press *Enter*

and you will go to the help file for the Stata `ttest` command

Use proper English and statistical terminology when doing a **Search**. Do not enter the name of a Stata command when doing a **Search**. Example:
- `t test` with **Search** is correct
- `ttest` with **Search** will work, but in general it is better to use proper English and statistical terminology with **Search**
- `ttest` in the Help window edit field is correct because `ttest` is a Stata command
- `ttest` with **Stata command...** is correct because `ttest` is a Stata command
- `t test` in the Help window edit field is incorrect

The command help files also contain hypertext links.
- Click on a hypertext link (shown by green words) and you will go to another help file.

When you have stepped through a series of help files:
- Click on **Back** to go back to the previous help file
- Click on **Top** to go back to the help table of contents or **Search** results screen

You can start again at any time:
- Just pull down **Help** from the main menu bar

Continued

Help, continued

The **Help** window has four buttons other than **Back** and **Top**:
  Search displays a dialog box where you may enter keywords
  to find help on a particular subject.

  Help! gives help and advice on using **Help**.

  Contents sends you to a subject table of contents with
  hypertext links for all help files.

  What's New describes the new features of Stata in this
  release or *Stata Technical Bulletin* update.

The help files contain lots of information.
  But not as much as the *Reference Manual*, *Graphics Manual*,
  and *User's Guide*.
  Help will tell you where to look in these manuals to find
  more information.

When we say "[U] **2.4 The Stata Technical Bulletin**", we mean
  section 2.4 in the *User's Guide*.

When we say "[R] **regress**",
  we mean the entry named **regress** in the *Reference Manual*.

When we say "[G] **graph options**",
  we mean the entry named **graph options** in the *Graphics Manual*.

The *Stata Technical Bulletin* (STB) provides
  • Updates to Stata—it's like getting a new release of Stata
    every other month
  • New programs written by us (StataCorp) and by our users
  • Articles on statistics and programming

The Stata web site (http://www.stata.com) provides
  • Answers to frequently asked questions (FAQs)
  • Links to STB FTP sites
  • Free additions to Stata ("Cool ado-files")
  • The latest official updates to Stata
  • Much, much more

The Help system

Let's illustrate the **Help** system with an example of how to use it.

1. *Choose* **Help** *from the main menu bar and select* **Search...**
2. *Enter* data *and click* **OK** *or press Enter.*

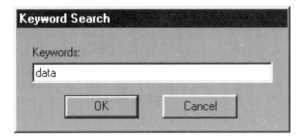

The Help system, continued

3. *Stata will now search for all references to "data" among the Stata commands, the Reference Manual, the Graphics Manual, the User's Guide, the Stata Technical Bulletin, and the FAQs on Stata's web site.* Here is the **Search** result:

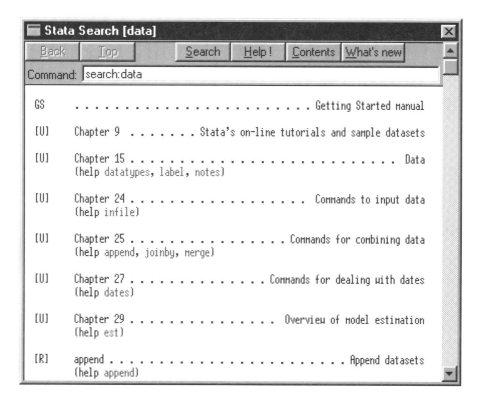

The Help system, continued

4. *Scroll down until you see* [R] describe.
   `describe` is a Stata command that will "Describe contents of data in memory or on disk". [R] means that the `describe` command is documented in the *Reference Manual*. The "help describe" means that there is an on-line help file for it.

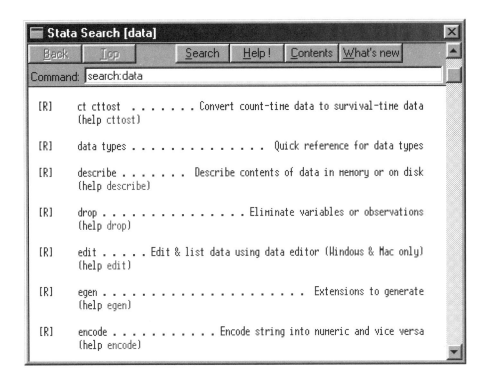

The Help system, continued

5. *Click on the hypertext link* "describe" *in* "help describe".
   Green words in the **Help** system are hypertext links. When you position the mouse pointer near a hypertext link, the pointer will change to a hand. Now when you click on the word, you will go to the corresponding help file. Clicking on the green "describe" takes you to the help file for the describe command.

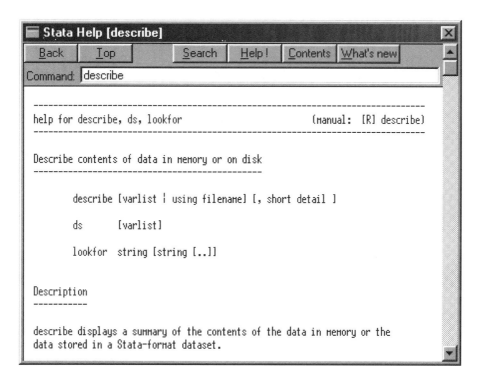

The Help system, continued

6. *Look at the help file for* describe.

   Help files for Stata commands contain (1) the command's syntax (i.e., the rules for how to type it), (2) a description of the command, (3) options, (4) examples, and (5) references to related commands. As you go through this manual, you will learn all about these things. For now, just scroll down to the examples. The examples tell us that it is valid to just type describe without anything else.

7. *Click on the Command window, type* describe, *and press Enter.*

   You may have to resize the Help window so that you can see the bottom of the Results window. We see that describe tells us that we have no observations.

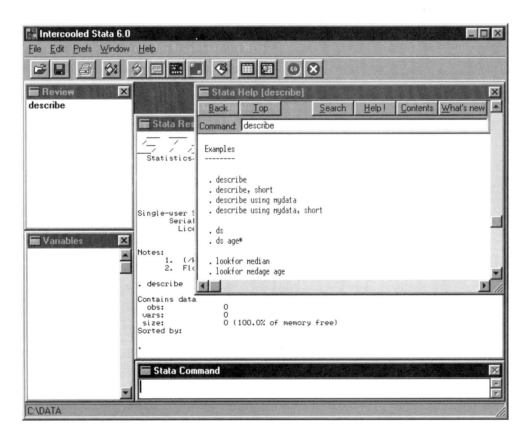

The Help system, continued

This example is the model of how we suggest you learn Stata and use Stata. To find out which Stata command will perform the statistical or data management task you would like to do:

1. Choose **Help**, select **Search...**, and enter the topic.

2. Scan the **Search** results, click on the command name hypertext link that is appropriate, and go to its help file.

3. With the help file open in front of you, click on the Command window and enter the command.

4. If the first help file you went to is not what you wanted, either look at the end of the help file for hypertext links to take you to related help files or click on **Top** to go back to the **Search** results and go from there to other help files.

5. If, at any time, you want to begin again with a new **Search**, click the **Search** button or select **Help** on the main menu bar and choose **Search...**.

Searching

If you enter `correlations` in the **Search** dialog box, Stata will search for all the sources of information that have to do with correlations.

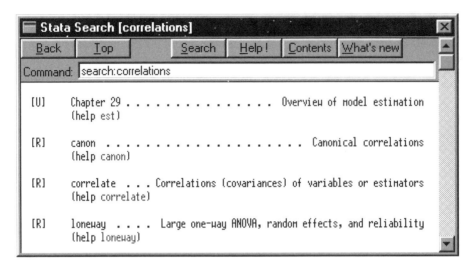

Search found several relevant entries in the *User's Guide* and *Reference Manual* and corresponding entries in the on-line help. If you are interested in correlations between variables, you could now click on `correlate`, to get a brief summary of the feature, or you could look at [R] **correlate** in the *Reference Manual* for a complete description.

Search also reported some relevant entries from the *Stata Technical Bulletin* (STB) but they are among the omitted output. See the end of this chapter for more information on these entries.

When you are using **Search**, use proper English and proper statistical terminology. Don't choose **Search...** and enter `ttest`. The term "ttest" is not proper statistical terminology. It is, however, the name of a Stata command. If you already know this and want to go directly to the help file for the `ttest` command, choose **Stata command...** from the **Help** menu, and type `ttest`. Or type `ttest` in the edit field at the top of the Help window and press *Enter*.

Help needs to distinguish between topics and Stata commands since some of the names of Stata commands are general topics. For example, `logistic` is a Stata command. If you choose **Stata command...** and enter `logistic`, you will go right to the help file for the command. But if you choose **Search...** and enter `logistic`, you will get search results listing the many Stata commands that relate to logistic regression.

Search is designed to help you find information about statistics, graphics, data management, and programming features in Stata. So when entering topics for the search, use appropriate terms from statistics, etc. For example, you could enter `Mann-Whitney`. Multiple topic words are allowed; e.g., `regression residuals`. For advice on using **Search**, click on the **Help!** button.

The Search, Help!, Contents, and What's New buttons

Four buttons appear at the top of a Help window: **Search, Help!**, **Contents**, and **What's New**.

Search allows you to perform a keyword search for help files.

Help! gives help for using **Help**.
It gives rules, advice, and examples for using the **Help** system. Especially valuable are its advice and examples for using **Search**. Take a look at it.

What's New describes the new features of Stata in this release or *Stata Technical Bulletin* update.

Contents contains a table of contents for all the command help files arranged by subject.

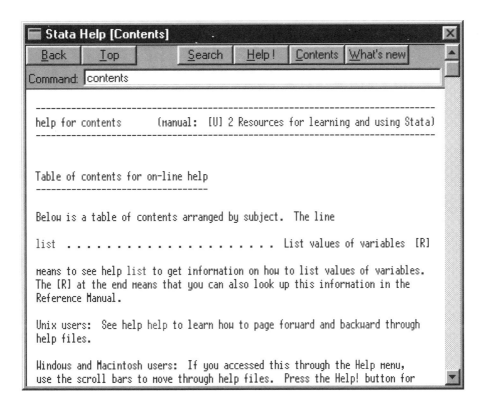

The Search, Help!, Contents, and What's New buttons, continued

There are over 200 entries in **Contents** and the output looks like

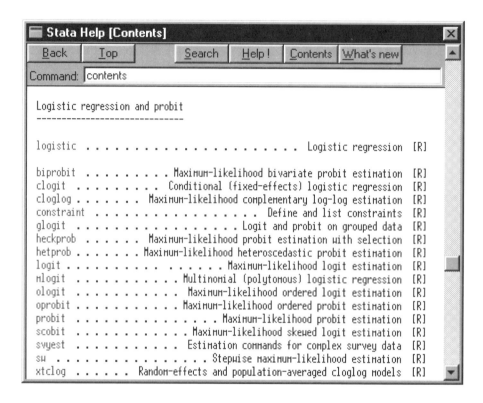

On the left is a hypertext link (almost always the name of a Stata command) that will take you to its help file. On the right is either '[R]' indicating that the *Reference Manual* also contains information on the topic, '[G]' indicating that the *Graphics Manual* contains information on the topic, '[U] *##*' indicating the section in the *User's Guide* where additional information can also be found, or '[GSW]' indicating that the *Getting Started with Stata for Windows* manual contains information on the topic.

help and search commands

Notes

1. *You can also access Stata's help system from the Command window.*
   When you do so, the output can appear either in the Results window or in the Help window.

2. *Typing* `search` *topic*
   in the Command window produces the same output as choosing **Search...** from the **Help** menu and entering *topic*. However, the output will appear in the Results window.

3. *Typing* `help` *commandname*
   gives the same output as choosing **Stata command...** from the **Help** menu and entering *command-name*, except the output will appear in the Results window.

4. **Important difference:**
   With the `help` and `search` commands, you will not have hypertext links in the Results window which you can use to jump to other help files.

5. *You can get hypertext help from the command line.*
   Rather than typing `help` *commandname*, type `whelp` *commandname*. The help file will appear in Stata's Help window, and you will be able to access the hypertext links. (Typing `whelp` *commandname* does the same thing as choosing **Stata command...** from the **Help** menu and entering *commandname*.)

6. See [U] **8 Stata's on-line help and search facilities** and [U] **8.8 search: All the details** in the *User's Guide* for more information about these command language versions of the **Help** system. The `search` command, in particular, has a few capabilities (such as author searches) that you cannot access via the **Help** dialog box.

The Stata Reference Manual, Graphics Manual, and User's Guide

Things like [R] **ci**, [R] **hotel**, and [R] **ttest** in the **Search** results and help files are references to the 4-volume *Stata Reference Manual*. You may also see things like [G] **printing Windows**, which are references to the *Stata Graphics Manual*, and things like [U] **12 The Break key**, which are references to the *Stata User's Guide*. Why bother to look in the manual when help is available on-line? Because the manual has more information.

Stata's on-line help is extensive. If you zoomed through it at one second per line, it would take you several hours to read it all. However, although the on-line help might seem complete, it contains only about one-tenth of the information contained in the *Reference Manual*.

The *Reference Manual* is like an encyclopedia; it is in straight alphabetical order. There are a few exceptions where some strongly related commands are grouped together. When looking for information on a command, it is always a good idea to start with the index to the manual.

Entries are named things like **collapse**, **egen**, ..., **summarize**, which are generally themselves Stata commands.

For some advice on how to use the *Reference Manual* see Chapter 18 in this manual, or see [U] **1.1 Getting Started with Stata**.

The Stata Technical Bulletin

The **Search** facility searches all the Stata source materials including the on-line help, the *User's Guide*, the *Reference Manual*, the *Graphics Manual*, this manual, the *Stata Technical Bulletin* (STB), and the FAQs on Stata's web site.

The STB, begun in May 1991, is a publication for and by Stata users—users of all disciplines and levels of sophistication. The STB contains articles written by us (StataCorp), Stata users, and others. Articles include descriptions of new and updated Stata commands (ado-files), tutorials on programming, illustrations of data analysis, discussions on teaching statistics, debates on appropriate statistical techniques, announcements, questions and answers, and suggestions.

Associated with each issue is an STB disk that contains the programs and datasets described in the STB. For instance, many of the new commands in this release were made available to STB subscribers long before the release itself.

Since the STB has had a number of articles dealing with correlations, if we choose **Search...** from the **Help** menu and enter `correlations`, we will see some STB references:

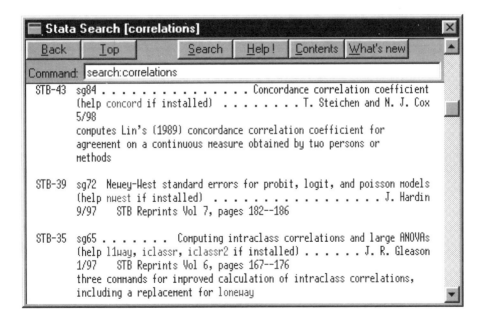

The Stata Technical Bulletin, continued

STB-39 refers to the thirty-ninth issue of the STB. Every six issues make a year's worth of STBs and they are bound and republished in a volume called *Stata Technical Bulletin Reprints*. STB-39 could thus be found in *STB Reprints*, Volume 7. Things like sg84, sg72, etc., are the way the STB refers to the articles that have appeared in it.

In the screenshot on the previous page, you will notice that article sg84 has a command `concord` associated with it. If you subscribed to the STB with disks at the time that STB-43 was published, you would have received a diskette with the `concord` command.

Links to sites where you can freely download the contents of each STB diskette can be found on the Stata web site. Point your browser to http://www.stata.com. Look for STB FTP under the user support section of the web site. You will find complete instructions to help you download and install the programs in the STB. If you use Stata for Windows 98/95/NT, you can download and install STB entries from within Stata. See Chapter 19 for more details.

We recommend that all users subscribe to the STB. See [U] **2.4 The Stata Technical Bulletin** for more information.

4 Inputting data with the data editor

The data editor

In this chapter, you will learn:

To enter the editor:
- Click on the **Data Editor** button
  (Windows 3.1 users click on the **Editor** button)
- Or type `edit` and press *Enter* in the Command window

The editor is like a spreadsheet.
   Columns correspond to variables and rows to observations.
- You navigate by clicking on a cell, or by using the arrow keys

You can copy and paste data between Stata's editor and other spreadsheets:
- Highlight the data you wish to copy in either Stata or the other spreadsheet and then pull down **Edit** and choose **Copy**
- After copying the data, you can paste it into Stata's editor or another spreadsheet by selecting the top left cell of the area to which you wish to paste, pulling down **Edit** and choosing **Paste**

You modify or enter data by
- Choosing the cell, typing the value, and pressing *Enter* or *Tab*

The difference between *Enter* and *Tab*:
- *Enter* moves you down to the next cell in the column
- *Tab* moves you to the right to the next cell in the row (until you get to the last filled-in column, then it takes you back to the first column)

To input data variable-by-variable:
- Click on the top cell in the first empty column
- Enter the value
- Press *Enter* to go down

To input data observation-by-observation:
- Click on the first cell in the first empty row
- Type the value
- Press *Tab* to go right
- After the first observation's values have been entered, click on the second cell in the first column
- Type the values for the second observation; press *Tab* to go right
- At the end of the second observation (and all subsequent ones), *Tab* will automatically take you back to the first column

---

Continued

The data editor, continued

Numeric data and string data (i.e., data consisting of characters) are entered the same way.
- You do not have to type double quotes around strings

Missing values for numeric variables are recorded as '.' (i.e., a period).
To enter a missing numeric value:
- Just press *Enter* or *Tab*
- Or type '.' and press *Enter* or *Tab*

Missing values for string variables are just empty strings (i.e., nothing).
To enter a missing string value:
- Just press *Enter* or *Tab*

The editor initially names variables `var1`, `var2`, ...
You can rename them.

To rename a variable:
- Double-click anywhere in the variable's column

This brings up the **Variable Information** dialog:
- Enter the new name of the variable

A variable name must be 1 to 8 characters long.
- The characters can be letters: `A – Z`, `a – z`
- Or digits: `0 – 9`
- Or underscores: `_`
- But no spaces or other characters

Example: `My_Name1`
The first character must be a letter or an underscore, but using an underscore to begin your names is not recommended.

Stata is case-sensitive.
- `Myvar`, `myvar`, and `MYVAR` are different names

Inputting data

Note to experienced Stata users: The data editor does what Stata's `input` command does (and much more).

Suppose we have the following data:

Make	Price	MPG	Weight	Gear Ratio
VW Rabbit	4697	25	1930	3.78
Olds 98	8814	21	4060	2.41
Chev. Monza	3667		2750	2.73
AMC Concord	4099	22	2930	3.58
Datsun 510	5079	24	2280	3.54
	5189	20	3280	2.93
Datsun 810	8129	21	2750	3.55

Note that we do not know MPG for the third car or the make of the sixth.

We will enter this dataset using the data editor.

1. *Call up the editor:*
   If you use Stata for Windows 98/95/NT, click on the **Data Editor** button. If you use Stata for Windows 3.1, click on the **Editor** button. Or type `edit` in the Command window.

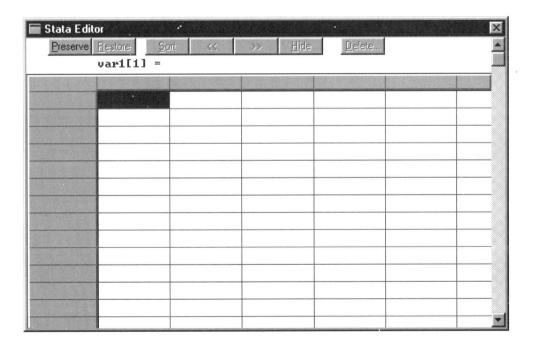

Inputting data, continued

2. *Enter the data.*
   Data can be entered variable-by-variable or observation-by-observation. Columns correspond to variables and rows to observations.

3. *When entering data observation-by-observation, press Tab after each value.*
   To input data observation-by-observation, start in the top cell of the first column. Type the make VW Rabbit and hit *Tab* to go right. You do not want to hit *Enter* since it moves you down.

   Now enter the price 4697 and hit *Tab*. Continue until you have entered all the values for the first observation. Now click on the second cell in the first column, and enter the values for the second observation using the *Tab* key.

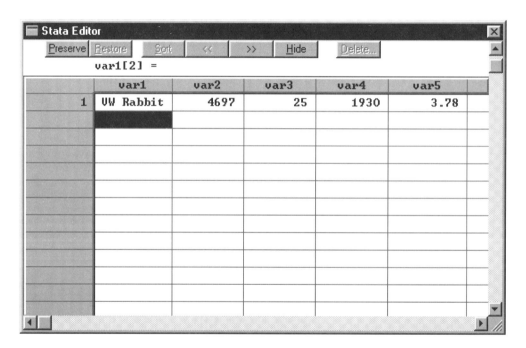

4. *The Tab key is smart.*
   After the first observation has been entered, Stata knows how many variables you have. So, at the end of the second observation (and all subsequent observations), *Tab* will automatically take you back to the first column.

5. *When entering data variable-by-variable, press Enter after each value.*
   To enter data variable-by-variable, click on the top cell in the first empty column. Type the values for the variable and press *Enter* after each one.

Inputting data, continued

Things to know about entering data

1. *Quotes around strings are unnecessary.*
   For other Stata commands, you will learn that you must type double quotes (") around strings. You can use double quotes in the editor, too, but you do not have to bother.

2. *A period ('.') represents a missing numeric value.*

3. *Just press Tab or Enter to input a missing numeric value.*
   MPG for the third observation is missing. To enter the missing value, just press *Tab* (or *Enter*). Or you could type '.'.

4. *Just press Tab or Enter to input a missing value for a string variable.*
   We do not know the make of the sixth car. So just press *Tab* (or *Enter*), and there will be an empty string (i.e., nothing) for the value of make for this observation.

5. *Stata will not allow empty columns or rows in the middle of your dataset.*
   Whenever you enter new variables or observations, always begin in the first empty column or row. If you skip over some columns or rows, Stata will fill in the intervening columns or rows with missing values.

6. *When you see, for example,* var3[4]=
   This corresponds to the current cell that is highlighted. var3 is the default name of the third variable, and [4] indicates the fourth observation.

Stata Editor window showing:

	var1	var2	var3	var4	var5
1	VW Rabbit	4697	25	1930	3.78
2	Olds 98	8814	21	4060	2.41
3	Chev. Monza	3667	.	2750	2.73
4	AMC Concord	4099	22	2930	3.58
5	Datsun 510	5079	24	2280	3.54
6		5189	20	3280	2.93
7	Datsun 810	8129	21	2750	3.55

var3[4] = 22

Renaming variables

1. *The editor initially names variables* var1, var2, ..., var5 *when it creates them.*

2. *To rename a variable, double-click anywhere in the variable's column.*
   Double-clicking on any cell in the variable's column brings up the **Variable Information** dialog. Enter the new name of the variable. (You can also enter a variable label; see Chapter 7. And you can change the display format of the variable; see [U] **15.5 Formats: controlling how data is displayed**.)

   We will call the first variable make, the second price, the third mpg, the fourth weight, and the fifth gratio. Just before we double-clicked and renamed var5 to gratio, our screen looked like

Stata Editor					×
Preserve Restore	Sort	<<	>>	Hide	Delete...
var5[3] = 2.73					

	make	price	mpg	weight	var5
1	VW Rabbit	4697	25	1930	3.78
2	Olds 98	8814	21	4060	2.41
3	Chev. Monza	3667	.	2750	2.73
4	AMC Concord	4099	22	2930	3.58
5	Datsun 510	5079	24	2280	3.54
6		5189	20	3280	2.93
7	Datsun 810	8129	21	2750	3.55

Rules for variable names

1. *Stata is case-sensitive.*
   Make, make, and MAKE are all different names to Stata. Had we called our variables Make, Price, MPG, etc., we would have to type them correctly capitalized in the future. Using all lowercase letters is easier.

2. *A variable name must be 1 to 8 characters long.*

3. *The characters can be letters (A–Z, a–z), digits (0–9), or underscores (_).*

4. *Spaces or other characters are not allowed.*

5. *The first character of a variable name must be a letter or an underscore.*
   But using an underscore to begin your names is not recommended, since Stata's built-in variables begin with an underscore.

Copying and pasting data

1. *Select the data you wish to copy.*

   a. Click once on a variable name to select the entire column.

   b. Click once on an observation number to select the entire row.

   c. Click and drag the mouse to select a range of cells.

   As an example, we will copy and paste an observation from Stata's editor into itself. Highlight the observation by clicking on the observation number.

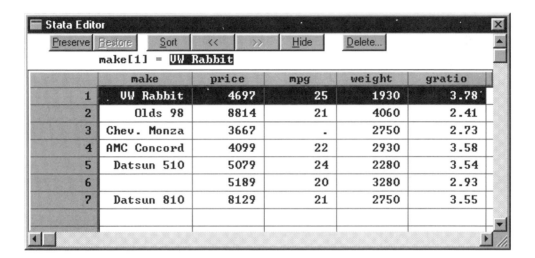

2. *Copy the data to the clipboard.*
   Pull down the **Edit** menu and choose **Copy**.

Copying and pasting data, continued

3. *Paste the data from the clipboard.*

   a. Click on the top left cell of the area to which you wish to paste.

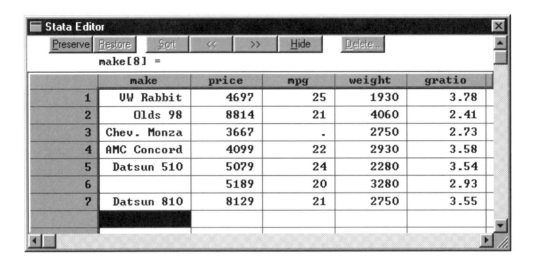

   b. Pull down the **Edit** menu and choose **Paste**.

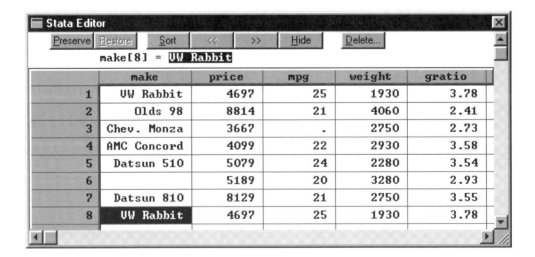

Exiting the data editor

1. *To exit the editor:*
   Click on the editor's close box (the box with an **X** at the right of the editor's title bar). If you are using Windows 3.1, double-click on the editor's control-menu box (the box at the left of the editor's title bar with the minus sign in it).

2. *Changes made in the editor are not saved until you tell Stata to save them.*
   The data that you have entered only exist in computer memory. They are not yet saved on disk.

3. *Save your data by pulling down* **File** *and choosing* **Save As**.
   We will enter the filename `afewcars` and Stata then adds the extension `.dta`, so the file is saved as `afewcars.dta`. See the next chapter for full details on saving your data.

4. *Note: You cannot save your data to disk until you exit the editor.*
   If you pull down the **File** menu while in the editor, **Save** and **Save As** will be grayed out.

Important implication: Changes are not permanent until you save them to disk. In Stata, you work with data in memory, not with data on a disk file. You can tell Stata to make a temporary backup before making changes; read about the **Preserve** button in Chapter 8.

Fonts and the data editor

Want to use a different font in the editor? A larger font? A smaller font? While in the editor, click on the editor's control-menu box, choose **Font**, and change it.

This change will affect the font in the editor this session, but not the next time you use Stata. To make the change permanent, pull down **Prefs** from Stata's main menu bar and choose **Save Windowing Preferences**.

You can change the fonts of any of Stata's windows in this way (except for the Command and Log windows). See Chapter 17 for details.

More on the data editor

The data editor also has a 'browse' mode which lets you safely look at your data without the possibility of accidentally changing it. See Chapter 8 for more information on the browse mode and other advanced features of the data editor.

Notes

5 Saving and loading data

How to save your data to disk and load it from disk

In this one-page chapter, you will learn:

To save new data (or old data under a new name):
- Pull down the **File** menu and choose **Save As**
- Or type `save` *filename* in the Command window

To resave data that has been changed (overwriting the original data file):
- Pull down the **File** menu and choose **Save**
- Stata for Windows 98/95/NT users may also click on the **Save** button
- Or type `save` *filename*`, replace`
- Or simply type `save, replace`

To load a data file:
- Pull down the **File** menu and choose **Open**
- Or type `use` *filename*
- Stata for Windows 98/95/NT users may also click on the **Open** button

When loading data, any data currently in memory is discarded.
Either save the current data first or allow it to be discarded:
- Choosing **File–Open** will ask for your **OK** before discarding it
- Or type `use` *filename*`, clear`
- Or type `clear` and then `use` *filename*

Stata data files are named *filename*`.dta`.
When saving or loading data files, the extension `.dta` is
automatically added when an extension is not specified.

Important note: Changes are not permanent until you save them. You work with a copy of the data in memory, not with the data file itself.

Comment about resaving data: There is no way to recover your original data file once you have done a **File–Save** or `save, replace`. With important datasets, you may want either to keep a backup copy of your original *filename*`.dta`, or to save the changed dataset under a new name by choosing **File–Save As** or by typing `save` *newfilename*.

Notes

6 Inputting data from a file

insheet, infile, and infix

In this chapter, you will learn:

Stata can read text (ASCII) files.	

There are 3 different commands for doing so:	`insheet`
	`infile`
	`infix`

If the text file was created by a spreadsheet or a database program:	Use `insheet`.
If the data in the text file is separated by spaces and does not have string variables (i.e., nonnumeric characters), or if all strings are just one word, or if all strings are enclosed in quotes:	Use `infile`.
Otherwise, you will have to specify the format of the data in the text file:	Use `infix` or `infile` (fixed format).
Spreadsheets save text files with delimiters:	`insheet` can read files delimited with commas or tab characters.
`insheet` cannot read space-delimited files:	Use `infile` as described on the next page.
`insheet` is easy to use:	It will read your variable names from the file, or it will make up variable names for you.
`insheet` is flexible:	You can override the variable names it would have chosen for you.
Before reading in data, you must clear memory. First save the current data (if you want it): Then clear memory by typing:	Choose **Save** from the **File** menu. `clear`

To read a text file created by a spreadsheet: To read `myfile.raw`: To read `myfile.xyz`: The extension `.raw` is assumed if you do not specify another extension.	`insheet using` *filename*
	`insheet using myfile`
	`insheet using myfile.xyz`

insheet, infile, and infix, continued

Suppose you have data for three numeric variables
    separated by spaces entered in a text file.
    To read them in with the names a, b, and c, type: `infile a b c using` *filename*
    To read from file `myfile.raw`:   `infile a b c using myfile`
    To read from file `myfile.xyz`:   `infile a b c using myfile.xyz`
    The extension `.raw` is assumed if you
    do not specify another extension.

Variables are given the datatype `float`
    unless you specify otherwise: a, b, c will be `float`s.

Stata has six datatypes for variables:

Real numbers, 8.5 digits of precision	`float`
Real numbers, 16.5 digits of precision	`double`

Integers between:

−127 and 126	`byte`
−32,768 and 32,766	`int`
−2,147,483,648 and 2,147,483,646	`long`

Strings (from 1 to 80 characters)

1-character long strings	`str1`
2-character long strings	`str2`
3-character long strings	`str3`
⋮	⋮
80-character long strings	`str80`

Continued

insheet, infile, and infix, continued

The numeric variable types `byte` and
   `int` are used when you want to
   reduce the memory that your dataset
   requires: See [U] **15 Data** and
                                     [U] **7 Setting the size of memory**.

The numeric variable types `long` and
   `double` are used for special
   accuracy requirements: See [U] **16.10 Precision and problems therein**.

You specify the datatype by putting it
   in front of the variable name.
   If you put nothing in front, the
   variable will be a `float`.

   To infile 3 numeric variables: `infile a b c using myfile`

   If string a is 10 characters or less
   and b and c numeric: `infile str10 a b c using myfile`

     If a is numeric, b a string, c numeric: `infile a str10 b c using myfile`

     If a and b are numeric, c a string: `infile a b str10 c using myfile`

     If a and b are strings, c numeric: `infile str10 a str10 b c using myfile`

     or: `infile str10 (a b) c using myfile`

Missing values:
   Missing numbers are indicated by: `.`
   Missing strings are indicated by: `""`

String values are in double quotes: `"Harry Smith"`
   If the string has no blanks, then
   you can omit the quotes: `Smith`

Stata can read formatted ASCII files.
   We give an example, but it is too
   complex to explain fully here: See [R] **infile (fixed format)**,
                                 [R] **infix (fixed format)**, and
                                 [U] **24 Commands to input data**.

Other programs can be purchased that
   convert other software formats
   to Stata format: See [U] **24.4 Transfer programs**.

insheet

The `insheet` command was specially developed to read in text (ASCII) files that were created by spreadsheet programs. Many people have data that they enter into their spreadsheet programs. All of the spreadsheet programs have an option to save data as a text (ASCII) file. The spreadsheet programs will delimit the columns in the text files with either tab characters or commas. Additionally, the spreadsheet programs will sometimes save the column titles (variable names in Stata) in the text file.

In order to read in this file, you only have to type `insheet using` *filename*, where *filename* is the name of the text file. The `insheet` command will determine whether there are variable names in the file, what the separator character is (tab or comma), and what type of data is in each column. If *filename* contains spaces, put double-quotes around the filename.

If you have a text (ASCII) file that you created by saving data from a spreadsheet program, then try the `insheet` command to read that data into Stata.

Remember that the `insheet` command only understands files that use the tab or comma as the column delimiter. If you have a file that uses spaces as the delimiter, use the `infile` command instead. See the `infile` section later in this chapter for more information.

insheet, continued

Suppose you have saved the file `sample.raw` from your favorite spreadsheet. It contains the lines:

```
VW Rabbit,4697,25,1930,3.78
Olds 98,8814,21,4060,2.41
Chev. Monza,3667,,2750,2.73
,4099,22,2930,3.58
Datsun 510,5079,24,2280,3.54
Buick Regal,5189,20,3280,2.93
Datsun 810,8129,,2750,3.55
```

These correspond to the make, price, MPG, weight, and gear ratio of a few cars. Note that the variable names are not in the file (so `insheet` will assign its own names) and the fields are separated by a comma. Use `insheet` to read the data:

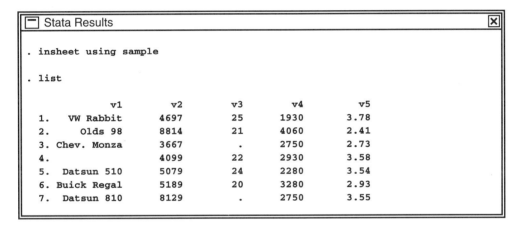

```
□ Stata Results                                               ☒

. insheet using sample

. list

            v1        v2      v3        v4        v5
  1.   VW Rabbit     4697      25      1930      3.78
  2.      Olds 98    8814      21      4060      2.41
  3. Chev. Monza     3667       .      2750      2.73
  4.                 4099      22      2930      3.58
  5.   Datsun 510    5079      24      2280      3.54
  6. Buick Regal     5189      20      3280      2.93
  7.   Datsun 810    8129       .      2750      3.55
```

If you want to specify better variable names, you can include those desired names in the command:

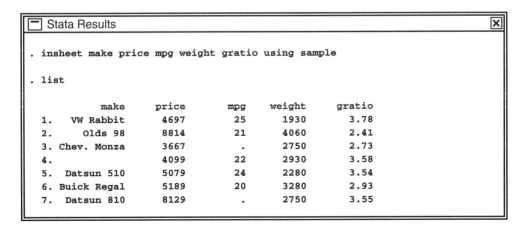

```
□ Stata Results                                               ☒

. insheet make price mpg weight gratio using sample

. list

            make     price     mpg    weight    gratio
  1.   VW Rabbit     4697      25      1930      3.78
  2.      Olds 98    8814      21      4060      2.41
  3. Chev. Monza     3667       .      2750      2.73
  4.                 4099      22      2930      3.58
  5.   Datsun 510    5079      24      2280      3.54
  6. Buick Regal     5189      20      3280      2.93
  7.   Datsun 810    8129       .      2750      3.55
```

infile

File `afewcars.raw` contains the lines:

```
"VW Rabbit"        4697        25        1930    3.78
"Olds 98"    8814    21   4060   2.41
"Chev. Monza"      3667        .          2750    2.73
""                 4099        22         2930    3.58
"Datsun 510"       5079        24         2280    3.54
"Buick Regal"
 5189
 20
 3280
 2.93
"Datsun 810"       8129       ***         2750    3.55
```

These correspond to the make, price, MPG, weight, and gear ratio of a few cars. Note, the second line is not formatted the same as the first line; we do not know MPG on the third line (file contains '.'); we do not know make on the fourth line (file contains ""); the sixth car's data is spread over 5 lines; we do not know MPG on the last line (file contains '***').

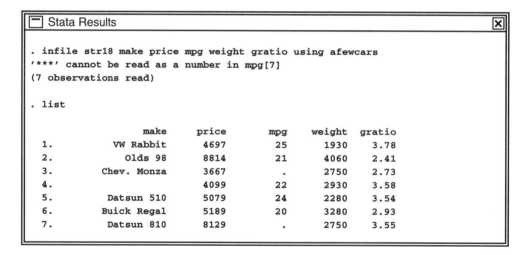

```
 ┌─┐ Stata Results                                                      ⊠

 . infile str18 make price mpg weight gratio using afewcars
 '***' cannot be read as a number in mpg[7]
 (7 observations read)

 . list

                     make      price      mpg     weight   gratio
         1.       VW Rabbit      4697       25       1930     3.78
         2.         Olds 98      8814       21       4060     2.41
         3.     Chev. Monza      3667        .       2750     2.73
         4.                      4099       22       2930     3.58
         5.     Datsun 510       5079       24       2280     3.54
         6.     Buick Regal      5189       20       3280     2.93
         7.     Datsun 810       8129        .       2750     3.55
```

Notes

1. using `afewcars` means using `afewcars.raw`. The extension `.raw` is assumed if you do not specify otherwise. If the data had been stored in `afewcars.asc`, we would have had to specify using `afewcars.asc`.

infile, continued

2. As in the data editor, '.' is understood to mean numeric missing value.

3. As in the data editor, an empty string ("") means string missing value. However, with `infile` you must explicitly have a "" in your file so that it does not read the next number in the file as the value of this string variable.

4. Observations need not be on a single line. The location of line breaks does not matter.

5. Things that are not understood (such as ***) are mentioned and stored as missing values.

infile with formatted data

File `cars2.raw` contains the lines:

```
VW Rabbit        4697      25      1930    3.78
Olds 98          8814      21      4060    2.41
Chev. Monza      3667              2750    2.73
                 4099      22      2930    3.58
Datsun 510       5079      24      2280    3.54
Buick Regal      5189      20      3280    2.93
Datsun 810       8129              2750    3.55
```

This data is more difficult to read because (1) there are no double quotes around the strings in the first column and they include blanks, and (2) there are blanks in the third column when we do not know the value.

`infile`'s ordinary logic when reading five variables is

1st thing	in the file is	1st variable	in 1st observation
2nd thing		2nd variable	in 1st observation
3rd thing		3rd variable	in 1st observation
4th thing		4th variable	in 1st observation
5th thing		5th variable	in 1st observation
6th thing		1st variable	in 2nd observation
7th thing		2nd variable	in 2nd observation
⋮		⋮ ⋮	

Carry out this logic on the above:

VW	is make	in 1st observation	
Rabbit	is price	in 1st observation	(error, stores as missing value)
4697	is MPG	in 1st observation	(wrong)
25	is weight	in 1st observation	(also wrong)
1930	is gear ratio	in 1st observation	(wrong again)
3.78	is make	in 2nd observation	(stores "3.78" as a string!)
Olds	is price	in 2nd observation	(error, stores as missing value)
98	is MPG	in 2nd observation	(surprise!)
⋮	⋮	⋮	

This problem is referred to as "loss of synchronization". There is a solution.

infile with formatted data, continued

Outside of Stata, using an editor or word processor, create `cars2.dct` containing the lines:

```
dictionary using cars2.raw {
        _column(1)    str18 make    %11s
        _column(19)   price         %4f
        _column(31)   mpg           %2f
        _column(39)   weight        %4f
        _column(47)   gratio        %4f
}
```

This is described in [R] **infile (fixed format)** in the *Reference Manual*. To read the data:

```
┌─────────────────────────────────────────────────────────────────────────┐
│ □ Stata Results                                                      [X] │
├─────────────────────────────────────────────────────────────────────────┤
│                                                                           │
│  . infile using cars2                                                     │
│                                                                           │
│  dictionary using cars2.raw {                                             │
│          _column(1)    str18 make    %11s                                 │
│          _column(19)   price         %4f                                  │
│          _column(31)   mpg           %2f                                  │
│          _column(39)   weight        %4f                                  │
│          _column(47)   gratio        %4f                                  │
│  }                                                                        │
│  (7 observations read)                                                    │
│                                                                           │
│  . list                                                                   │
│                                                                           │
│                     make      price      mpg    weight    gratio          │
│       1.        VW Rabbit      4697       25      1930      3.78          │
│       2.          Olds 98      8814       21      4060      2.41          │
│       3.      Chev. Monza      3667        .      2750      2.73          │
│       4.                       4099       22      2930      3.58          │
│       5.        Datsun 510     5079       24      2280      3.54          │
│       6.      Buick Regal      5189       20      3280      2.93          │
│       7.        Datsun 810     8129        .      2750      3.55          │
│                                                                           │
└─────────────────────────────────────────────────────────────────────────┘
```

That is,

1. Create another file describing the contents of the data with another program—an editor or word processor. Name the file *myfile*.`dct`, and save it as a text (ASCII) file. Note that editors such as Notepad and WordPad may add a .`txt` extension to the file, resulting in *myfile*.`dct.txt`. To prevent this, enclose the filename in double quotes (") when you save it from an editor.

2. Instructions for creating the contents of this file can be found in [R] **infile (fixed format)** of the *Reference Manual*.

3. Once the dictionary exists, you type `infile using` *myfile*. The file extension .`dct` is assumed when `infile` is used without any variable names.

Importing files from other software

If you have a file created by another software package that you would like to read into Stata, you should be able to use options on that software package to output the file as a plain text (ASCII) file. Then you can use `infile`, `insheet`, or `infix` to read in the data.

Or you may be able to copy and paste data from another spreadsheet into Stata's data editor. See [R] **edit** for full details.

Or you can purchase a transfer program that will convert the other software's data file format to Stata's data file format. See [U] **24.4 Transfer programs** in the *User's Guide*.

7　Labeling data

describe and label

In this chapter, you will learn:

To obtain a description of the data	
currently in memory:	`describe`
stored on disk:	`describe using` *filename*
To set or reset the data label:	`label data "`*text*`"`
To set or reset a variable's label:	`label var` *varname* `"`*text*`"`
To label a variable's values:	
Create a value label:	`label define` *lblname* `#` `"`*text*`"` `#` `"`*text*`"` ...
Attach value label to variable:	`label values` *varname* *lblname*
You can attach the same value label to other variables, too.	
To unlabel	
the data:	`label data`
a variable:	`label var` *varname*
a variable's values:	`label values` *varname*
To delete a value label:	`label drop` *lblname*
To change a value label:	
Delete it:	`label drop` *lblname*
Redefine it:	`label define` *lblname* `#` `"`*text*`"` `#` `"`*text*`"` ...
To make changes permanent,	
resave the data:	Choose **Save** under the **File** menu.
or, alternatively, you can type:	`save` *filename*`,` `replace`

Note

We cover only the rudiments of value labels here. For instance, there is a way to modify value labels without deleting and redefining. See [U] **15.6.3 Value labels** in the *User's Guide* for a complete treatment.

describe

At the end of Chapter 4, we saved a dataset called `afewcars.dta`:

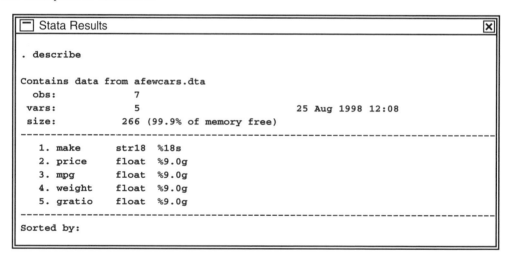

```
┌─────────────────────────────────────────────────────────────────────────────┐
│ ⊟ Stata Results                                                          ☒  │
├─────────────────────────────────────────────────────────────────────────────┤
│                                                                             │
│  . use afewcars                                                             │
│                                                                             │
│  . list                                                                     │
│                    make      price      mpg     weight     gratio          │
│        1.       VW Rabbit     4697       25       1930       3.78          │
│        2.         Olds 98     8814       21       4060       2.41          │
│        3.     Chev. Monza     3667        .       2750       2.73          │
│        4.     AMC Concord     4099       22       2930       3.58          │
│        5.      Datsun 510     5079       24       2280       3.54          │
│        6.                     5189       20       3280       2.93          │
│        7.      Datsun 810     8129       21       2750       3.55          │
│                                                                             │
└─────────────────────────────────────────────────────────────────────────────┘
```

The description of this data is

```
┌─────────────────────────────────────────────────────────────────────────────┐
│ ⊟ Stata Results                                                          ☒  │
├─────────────────────────────────────────────────────────────────────────────┤
│                                                                             │
│  . describe                                                                 │
│                                                                             │
│  Contains data from afewcars.dta                                            │
│    obs:            7                                                         │
│    vars:           5                          25 Aug 1998 12:08             │
│    size:         266 (99.9% of memory free)                                 │
│  ─────────────────────────────────────────────────────────────────────     │
│      1. make      str18    %18s                                             │
│      2. price     float    %9.0g                                            │
│      3. mpg       float    %9.0g                                            │
│      4. weight    float    %9.0g                                            │
│      5. gratio    float    %9.0g                                            │
│  ─────────────────────────────────────────────────────────────────────     │
│  Sorted by:                                                                 │
│                                                                             │
└─────────────────────────────────────────────────────────────────────────────┘
```

      variable name   display format

            ↘   ↙

     1. make   str18   %18s

                  ↑

          storage type

1. The variable name is how we refer to the column of data.

2. The storage type was described in Chapter 6.

3. The display format is described in [U] **15.5 Formats: controlling how data is displayed** of the *User's Guide*. It controls how the variable is displayed. By default, Stata sets it to something reasonable given the storage type.

Describing a file without loading it

It is not necessary to load a dataset before describing it:

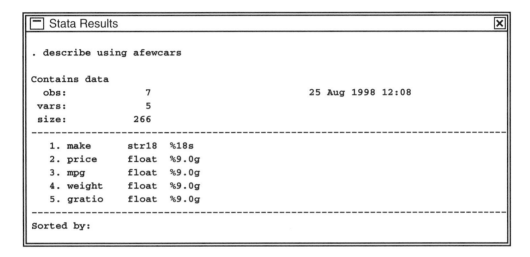

That is,

1. When you type describe by itself, Stata describes the current contents of memory.

2. When you type describe using *filename*, Stata describes the contents of the specified Stata data file (i.e., a file named *filename*.dta created by Stata).

describe and labels

Datasets can contain labels on the data, variables, and values. As an example, we use the dataset auto.dta which is stored in the directory where you installed Stata:

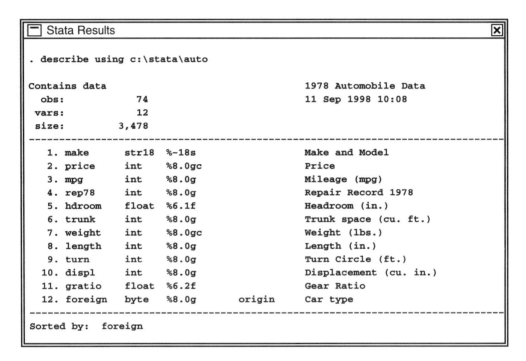

That is,

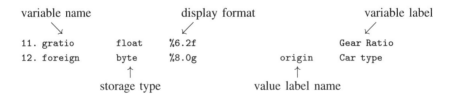

1. Value labels allow numeric variables—such as foreign—to have words associated with numeric codes. The describe tells us that the numeric variable foreign has value label origin associated with it. Although not revealed by describe, the variable foreign takes on the values 0 and 1 and the value label origin associates 0 ↔ Domestic and 1 ↔ Foreign. When we browse the data (see Chapter 8), it would appear as if foreign contained the values "Domestic" and "Foreign". Note that it is not necessary for the value label to have a name different than the variable. We could just as easily have used a value label named foreign.

2. Variable labels are merely comments that help us remember the contents of variables.

Labeling datasets and variables

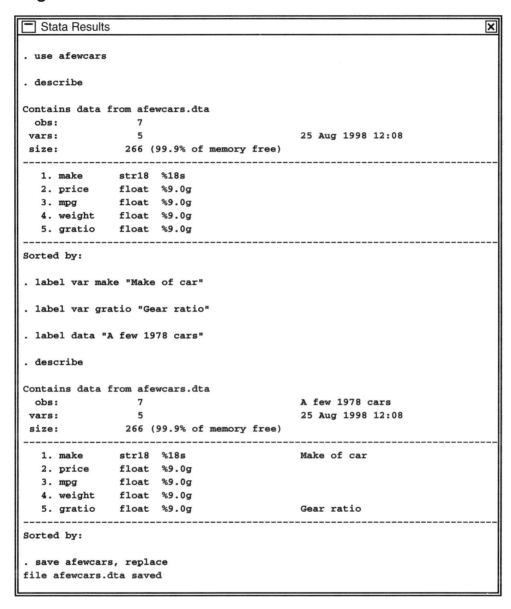

That is,

1. Use `label var` to set or reset a variable's label. Put the label itself in double quotes:

   `label var gratio "Gear ratio"`

2. Use `label data` to set or reset the dataset's label.

3. You set or change the labels on the data in memory. To make the changes permanent, resave the data.

Labeling values of variables

We have secretly added the variable source to afewcars.dta:

```
┌─ Stata Results ──────────────────────────────────────────────────────── ⊠ ┐
│                                                                            │
│  . list                                                                    │
│                                                                            │
│                    make      price       mpg     weight     gratio   source│
│      1.       VW Rabbit       4697        25       1930       3.78        1 │
│      2.         Olds 98       8814        21       4060       2.41        0 │
│      3.     Chev. Monza       3667         .       2750       2.73        0 │
│      4.     AMC Concord       4099        22       2930       3.58        0 │
│      5.      Datsun 510       5079        24       2280       3.54        1 │
│      6.                       5189        20       3280       2.93        0 │
│      7.      Datsun 810       8129        21       2750       3.55        1 │
│                                                                            │
└────────────────────────────────────────────────────────────────────────┘
```

Labeling values of variables, continued

source $= 0$ denotes domestic cars and 1 denotes foreign. Label the values "domestic" and "foreign":

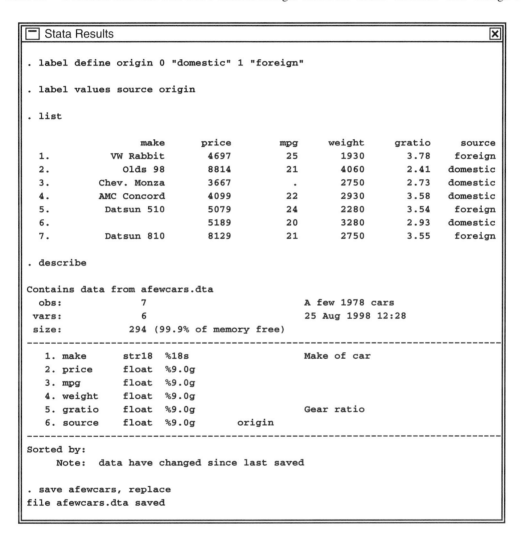

That is,

1. Use label define to create a value label. The syntax is
   label define *labelname* # *"contents"* # *"contents"* ...

2. Use label values to associate the label with a variable. The syntax is
   label values *variablename labelname*

3. To make the change permanent, resave the data.

Notes

8 Changing and viewing data with the data editor

Advanced use of the data editor

In this chapter, you will learn:

You can select the variables that appear in the editor:

Type in the Command window:

- `edit make`   Selects the single variable `make`
- `edit make mpg`   Selects the variables `make` and `mpg`

You can include any number of variables.

- See Chapter 9 for shortcuts for entering variable names.

You can restrict the observations that appear in the editor:

Type in the Command window:

- `edit in 1`   Uses only the first observation
- `edit in 2`   Uses only the second observation
- `edit in -2`   Uses only the second from the last observation
- `edit in -1`   Uses only the last observation
- `edit in l` (i.e., ℓ)   Also uses only the last observation

To restrict to a series of observations:

- `edit in 1/9`   Uses observations 1 through 9
- `edit in 2/-2`   Uses observations 2 through second from the last

To restrict to those observations satisfying a mathematical expression:

- `edit if` *exp*   Uses observations for which *exp* is true
- `edit if mpg>20`   Uses observations for which `mpg>20`
- `edit if mpg==20`   Uses observations for which `mpg` equals 20
- `edit if rep78==.`   Uses observations for which `rep78` equals missing

You can combine `in` and `if` (the order does not matter):

- `edit in 1/9 if mpg>=25`
- `edit if price<6000 in 5/-1`

You can select variables and restrict observations at the same time:

- `edit make in 5/-5`
- `edit make mpg if mpg>=25`
- `edit make mpg if rep78==.`
- `edit make mpg in 1/9 if mpg>=25`

Continued

Advanced use of the data editor, continued

To make changes to your data:
- Type `edit` alone or `edit` *varname(s)*, `edit if` ... , etc.
- Or click on **Data Editor** (but you cannot select variables or restrict observations if you enter the editor this way) (Windows 3.1 users click on the **Editor** button)
- Once in the editor, click on a cell you want to change
- Enter a new value and press *Enter* or *Tab*

Restricting the editor to certain variables and observations lessens the chance of making a serious mistake. For example, if you need to change `mpg` when it is missing, you can limit the data's exposure by typing
- `edit make mpg if mpg==.`

Note: For making systematic global changes to your data, the `replace` command may be more appropriate.
- See Chapter 11

To delete variables and observations:
- Press the **Delete...** button; see the next page

Note: For deleting groups of observations or variables, the `drop` command is more appropriate.
- See Chapter 12

The data editor can be used to view data.
To enter the editor in `browse` mode:
- Click on the **Data Browser** button (Windows 3.1 users click on the **Browse** button)
- Or type `browse` in the Command window

The `browse` mode is useful since you cannot accidentally change your data. Use `browse`, not `edit`, when you just want to look.

You can also select variables and restrict observations with `browse`.
Examples:
- `browse make mpg`
- `browse in 1/20`
- `browse if rep78==.`
- `browse make mpg rep78 in 5/-5 if mpg>20`

In later chapters, you will learn:

What is true for `edit` and `browse` is true for almost all Stata commands: Type the command optionally followed by a variable list optionally followed by an `if` optionally followed by an `in`.

Buttons on the data editor

The editor has seven buttons:

Preserve
> If you make changes to your data in the data editor and
> are satisfied with them (and plan to stay in the editor to
> make additional changes), you can update the backup copy:
> > • Click on **Preserve**

Restore
> Stata automatically makes a backup copy of your data
> when you enter the editor.
>
> If you want to cancel the changes that you made and
> restore the backup copy:
> > • Click on **Restore**

Sort
> **Sort** sorts the observations in ascending order of the current variable.

<<
> The << button shifts the current variable to be the first variable.

>>
> The >> button shifts the current variable to be the last variable.

Hide
> **Hide** hides the current variable.
> The variable still exists; the editor just stops displaying it.

Delete...
> **Delete...** brings up a dialog that allows you to
> > • Delete the current variable
> > • Delete the current observation
> > • Delete all observations throughout the dataset that have the
> >   same value for the current variable as the current observation

Changing data

1. *Load in the* `auto.dta` *dataset that we provide with Stata:*
   Pull down **File** and choose **Open**, and select `auto.dta` from the `c:\stata` directory (or from wherever Stata is installed).

2. *Enter the editor:*
   Click on the **Data Editor** button or type `edit` in the Command window.

3. *The* **Sort** *button sorts the observations in ascending order of the current variable.*
   Suppose we want to look at the lowest and highest priced cars. Click once anywhere on the `price` column, and then click on **Sort**. Scroll down to the bottom of the dataset to see the highest priced cars.

4. *The shift button* << (>>) *shifts the current variable to be the first variable (last variable).*
   Let's find the lightest and heaviest cars. Unless you have a big monitor, you will not be able to view both `make` and `weight` together. So scroll right, click on the `weight` column, and then click on <<. Now scroll left, select `weight`, and click on **Sort**. Select `make` and click on <<.

Stata Editor				
Preserve	Restore	Sort	<<	>>	Hide	Delete...

make[1] = Honda Civic

	make	weight	price	mpg
1	Honda Civic	1,760	4,499	28
2	Ford Fiesta	1,800	4,389	28
3	Plym. Champ	1,800	4,425	34
4	Renault Le Car	1,830	3,895	26
5	VW Rabbit	1,930	4,697	25
6	Mazda GLC	1,980	3,995	30
7	VW Scirocco	1,990	6,850	25
8	Datsun 210	2,020	4,589	35
9	VW Diesel	2,040	5,397	41
10	Subaru	2,050	3,798	35
11	Audi Fox	2,070	6,295	23
12	Chev. Chevette	2,110	3,299	29
13	Dodge Colt	2,120	3,984	30
14	Fiat Strada	2,130	4,296	21
15	VW Dasher	2,160	7,140	23
16	Plym. Horizon	2,200	4,482	25

Changing data, continued

5. *The* **Hide** *button hides the current variable.*
   The effect of **Hide** is only cosmetic. The variable still exists; it is simply not displayed. Click on the `rep78` column and click on **Hide**.

6. *To change a value, click on the cell, enter the new value, and press Enter or Tab.*
   Try doing this.

7. **Delete...** *allows you to delete one variable, one observation, or all observations that have a certain value.* Select the `trunk` variable. Press **Delete...** and choose **OK**. We dropped `trunk` from the dataset. This change is real; you cannot get `trunk` back unless you cancel all your changes. (But remember, no changes are permanent until you save them on disk. You work with a copy of the data in memory, not with the data file itself.)

8. *Choose* **Restore** *to restore the backup copy of your data.*
   Click on **Restore**. The dataset is now exactly as it was when you began. A backup copy of your dataset is automatically made when you enter the editor. (But this can be changed; see [R] **edit** in the *Reference Manual.*)

9. **Preserve** *updates the backup copy.*
   Sort again by `mpg`. Click on one of the cells that has `mpg` equal to 14. Press **Delete...**, select the third choice "Delete all 6 obs. where mpg==14", and click on **OK**.

   Click on **Preserve**. Now make some other changes to the data. Click on **Restore**. The changes that you made after the **Preserve** are reversed. There is no way to cancel the changes you made before you pressed **Preserve** (except to reload the original data file).

10. *When you exit the editor, a dialog box will ask you to confirm your changes.*
    If you **OK** your changes, the editor discards the backup copy.

Changing data, continued

11. *You will find output in the Results window documenting the changes you made.*
    What you see are Stata commands that are equivalent to what you did in the editor. The dash in
    front of the command indicates that the change was done in the editor.

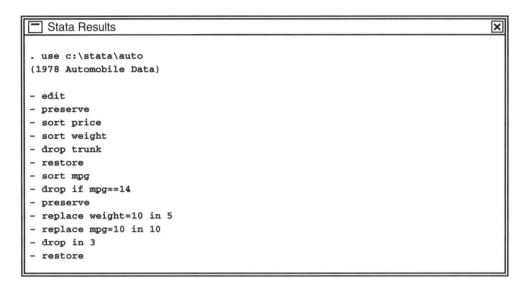

```
┌─┬──────────────────────────────────────────────────────────┬──┐
│ ☐ │ Stata Results                                          │ ☒ │
├─┴──────────────────────────────────────────────────────────┴──┤
│                                                                │
│  . use c:\stata\auto                                           │
│  (1978 Automobile Data)                                        │
│                                                                │
│  - edit                                                        │
│  - preserve                                                    │
│  - sort price                                                  │
│  - sort weight                                                 │
│  - drop trunk                                                  │
│  - restore                                                     │
│  - sort mpg                                                    │
│  - drop if mpg==14                                             │
│  - preserve                                                    │
│  - replace weight=10 in 5                                      │
│  - replace mpg=10 in 10                                        │
│  - drop in 3                                                   │
│  - restore                                                     │
│                                                                │
└────────────────────────────────────────────────────────────────┘
```

edit with in and if

Here is one example of using `edit` with selected variables and restricted observations. Chapter 10 contains many examples of using a command with a variable list and `in` and `if`.

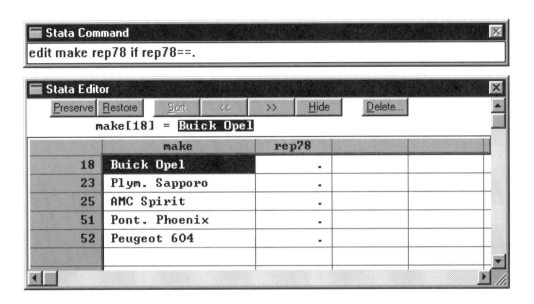

Note: You cannot use the **Sort** button if you entered the editor with an `in` or `if` restriction.

Advice

1. *People who care about data integrity know editors are dangerous—it is easy to accidentally make changes.* Never use `edit` when you just want to look at your data. Use `browse`.

2. *Protect yourself when you `edit` data by limiting the data's exposure.*
   If you need to change `rep78` only if it is missing, and need to see `make` to make the change, do not press the **Data Editor** button, but type

       edit make rep78 if rep78==.

   It is now impossible for you to change (damage) variables other than `make` and `rep78` and observations other than those with `rep78` equal to missing.

3. *All of this said, Stata's editor is safer than most because it records changes in the Results window.* Use this feature—log your output and make a permanent record of the changes. Then you can verify that the changes you made are the changes you wanted to make. See Chapter 16 for how to create a log file.

browse

1. *When you want to look at your data, but do not want to change it, you can enter the editor in* browse *mode:* Either click on the **Data Browser** button or type browse in the Command window.

2. *In* browse *mode, it is impossible to alter your data.*

3. *You can select variables and restrict observations with* browse, *just like* edit.

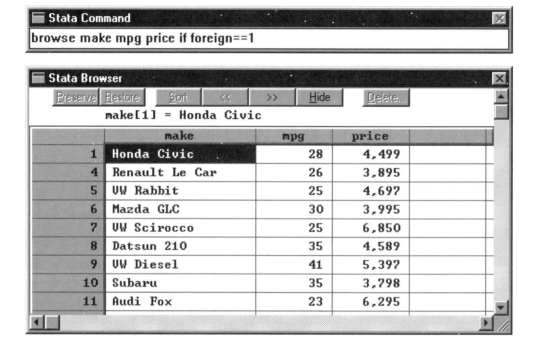

4. *You can use the* **Sort**, *shift* (<< *and* >>), *and* **Hide** *buttons in* browse *mode.*
   However, just like edit, you cannot sort if you entered browse with an in or if restriction.

5. browse *can be used to do much of what the* list *command does.* But because you can scroll in the data editor, the editor is more convenient for viewing many variables and for scrolling up and down through the observations.

   The list command, which produces output in the Results window, is described in Chapter 10. list is useful for producing listings for your log file (see Chapter 16) and for viewing a few variables quickly in the Results window.

9 Shortcuts: The Review and Variables windows

Entering commands quickly

In this chapter, you will learn:

The Review window shows your past commands.
    Click once on a past command and it is copied to the Command window.
    Double-click and the command is copied and executed.
    To see commands further back resize the Review window and make it longer.
    You can scroll this window if you have Stata for Windows 98/95/NT.
    You cannot scroll this window if you have Stata for Windows 3.1.
        • To view all previous commands and output, first create a log file
        • The Log window allows you to scroll; see Chapter 16

The Variables window shows the current variables.
    Click once on a variable and its name is copied to the Command window.
    (If you double-click, the variable will be copied twice.)
    You can scroll this window.

The Command window follows standard Windows editing style.

Keys for editing in the Command window:

Delete	Deletes the character to the right of the cursor
Backspace	Deletes the character to the left of the cursor
Esc	Deletes the entire command line
Home	Moves cursor to the beginning of the line
End	Moves cursor to the end of the line
Page Up	Retrieves previous command lines
Page Down	Steps forward through past command lines

(The *Page Up* and *Page Down* keys are equivalent to clicking on past commands in the Review window.)

You can copy from the Results and Log windows into the
    clipboard and paste into the Command window.
    If the clipboard contains more than one line, only
    the first line is pasted into the Command window.

You can also copy from the Graph window to the clipboard.
        • See Chapter 15

The do-file editor allows you to submit several commands at once to Stata.
        • See Chapter 14

Continued

More shortcuts

Many Stata commands can be abbreviated.
  Blue highlighting in the syntax diagrams in the command help files
  indicates the shortest allowable abbreviation.

  Underlining in the syntax diagrams in the *Reference Manual* also indicates
  the shortest allowable abbreviation.

  Variable names can also be abbreviated in many cases.
  - See [U] **14.2 Abbreviation rules** in the *User's Guide* for rules and full
    details (this manual also gives many examples)

The *F*-keys can be defined to issue commands or to give a list of variable names.
  For example, the *F3* key comes defined as `describe` *Enter*.
  You can easily redefine the *F*-keys to make your own customized shortcuts.

You can resize and rearrange any of Stata's windows.
  And you can save your arrangement or return to the default.
  - See Chapter 17

You can close the Review and Variables windows,
  but you cannot close the Command and Results windows.
  - See Chapter 17 for details about opening windows, etc.

You can change fonts in most of the windows:
  - See Chapter 17

Keyboard shortcuts for menus:
  These are standard:
  - Press *Alt* and the underlined letter of the menu name
  - E.g., for **F**ile, press *Alt-F*

Entering commands quickly

1. *Let's say that you have issued many different commands in a session working with Stata.*
   Suppose that one of these commands was a `regress` command, and now you want to add another variable to the regression and rerun it.

2. *Use the mouse and move the pointer into the Review window and over the `regress` command.*
   Click once. The previous command will be copied into the Command window.

3. *Move the pointer into the Variables window and click once on `foreign`.*
   The Command window now contains the command you want. Press *Enter* to issue the command.

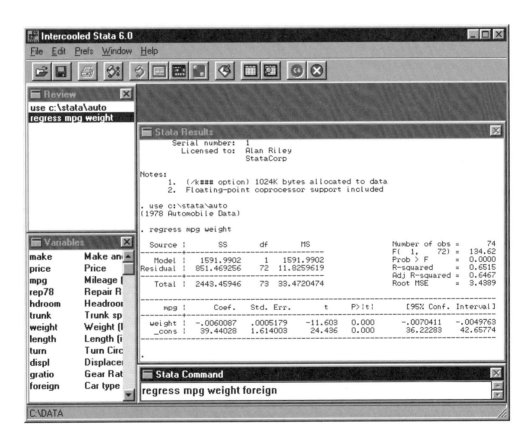

Entering commands quickly, continued

4. *If you want to reissue a previous command exactly as typed:*
   Just double-click on the command in the Review window. Double-clicking instantly executes the command. It is equivalent to copying the command into the Command window and pressing *Enter.*

5. *You can also use the Review window to quickly fix typos.*
   If you make a typo and issue the command without realizing it, you don not have to retype the whole command line again. Just go to the Review window, click once on the errant command to copy it into the Command window, then fix the typo, and hit *Enter.*

F-keys

1. *Some of the F-keys are defined to have special meanings.*
   For example, pressing *F3* is equivalent to typing `describe` and pressing *Enter.* Pressing *F7* is the same as typing `save` with a space after it—there is no *Enter* so you can now add the name of the file.

2. *Note: The default definitions of these keys are the same for all versions of Stata—Windows, Macintosh, and Unix.* For Stata for Windows users, the definitions are not particularly useful. However, this really does not matter, because the keys only take on great utility when you define them based on the problem at hand. To see the default definitions, type `macro list`.

3. *How is it that pressing F3 yields the same result as typing* `describe`?
   When Stata comes up, it silently issues the command

   ```
   global F3 "describe;"
   ```

   to itself. In this context, a semicolon means *Enter.* This makes the *F3* key equivalent to typing `describe` and then pressing *Enter.*

4. *To define the F-keys for yourself, you also use the* `global` *command.*
   For example, let's say you plan to run many `logistic` commands, and you don't want to have to type all those letters in `logistic` each time. If you type

   ```
   global F8 "logistic "
   ```

   then hitting *F8* is equivalent to typing `logistic` and a space.

5. *Suppose* all the logistic regressions that you plan to run will have the dependent variable `outcome` and will always include the three independent variables `drug`, `sex`, and `age`. And you want to play around with adding other variables to the regression. You can type

   ```
   global F9 "logistic outcome drug sex age "
   ```

   Now, each time you run a regression, you will only have to press *F9*, then go to the Variables window, click on the names of any additional independent variables, and press *Enter.*

6. *See* [U] **13 Keyboard use** *and* [U] **21.3 Macros** *in the User's Guide for more information.*

Notes

10 Listing data

list

In this chapter, you will learn:

`list` is similar to `browse`.	
To list in the Results window, type:	`list`
When a `--more--` appears in the Results window: (as happens when `list`ing a long list)	
To see the next line:	Press *Enter*.
To see the next screen:	Press any key, such as *Space Bar*.
or:	Click on the **More** button.
To interrupt a Stata command at any time and return to the state before you issued the command:	Click on the **Break** button.
or:	Press *Ctrl-Break*.
To list a single variable:	`list` *varname*
Example:	`list weight`
You may abbreviate `list`:	`l weight`
You may abbreviate *varname*:	`list wei`
To list any number of variables:	`list` *varname(s)*
Example:	`list make mpg`
You may abbreviate:	`l mak mpg`
To list *varname$_i$* through *varname$_j$*:	`list` *varname$_i$*-*varname$_j$*
Example:	`list make-mpg`
You may abbreviate:	`l mak-mpg`
To list all variables starting with `pop`:	`list pop*`
You may combine all the above:	`list mak-mpg wei pop*`

Continued

list, continued

To list the third observation:	`list in 3`
the second from last:	`list in -2`
the last:	`list in -1`
or:	`list in l` (i.e., the letter ℓ)
To list observations 1 through 3:	`list in 1/3`
5 to 17:	`list in 5/17`
3 to third from last:	`list in 3/-3`
You may combine all the above:	`list mpg pop* in 3/-3`
To conditionally list observations:	`list if` *exp*
Example:	`list if mpg>20`
You may combine all the above:	`list mpg wei pop* if mpg>20`
	`l mpg wei pop* if mpg>20 in 3/-3`
Output from `list` can be logged (as can all output that appears in the Results window):	See Chapter 16.

list with variable list

```
┌─ Stata Results ──────────────────────────────────────────────────────────[X]─┐
│                                                                               │
│  . list                                                                       │
│                                                                               │
│                    make      price       mpg     weight    gratio     source  │
│       1.       VW Rabbit       4697        25       1930      3.78    foreign  │
│       2.         Olds 98       8814        21       4060      2.41   domestic  │
│       3.     Chev. Monza       3667         .       2750      2.73   domestic  │
│       4.     AMC Concord       4099        22       2930      3.58   domestic  │
│       5.      Datsun 510       5079        24       2280      3.54    foreign  │
│       6.                       5189        20       3280      2.93   domestic  │
│       7.      Datsun 810       8129        21       2750      3.55    foreign  │
│                                                                               │
│  . list make mpg price                                                        │
│                                                                               │
│                    make       mpg      price                                  │
│       1.       VW Rabbit        25       4697                                 │
│       2.         Olds 98        21       8814                                 │
│       3.     Chev. Monza         .       3667                                 │
│       4.     AMC Concord        22       4099                                 │
│       5.      Datsun 510        24       5079                                 │
│       6.                        20       5189                                 │
│       7.      Datsun 810        21       8129                                 │
│                                                                               │
│  . list m*                                                                    │
│                                                                               │
│                    make       mpg                                             │
│       1.       VW Rabbit        25                                            │
│       2.         Olds 98        21                                            │
│       3.     Chev. Monza         .                                            │
│       4.     AMC Concord        22                                            │
│       5.      Datsun 510        24                                            │
│       6.                        20                                            │
│       7.      Datsun 810        21                                            │
│                                                                               │
│  . l price-weight                                                             │
│                                                                               │
│                   price       mpg     weight                                  │
│       1.          4697        25       1930                                   │
│       2.          8814        21       4060                                   │
│       3.          3667         .       2750                                   │
│       4.          4099        22       2930                                   │
│       5.          5079        24       2280                                   │
│       6.          5189        20       3280                                   │
│       7.          8129        21       2750                                   │
│                                                                               │
└───────────────────────────────────────────────────────────────────────────┘
```

list with variable list, continued

Notes

1. `list` without arguments lists the entire dataset. You can always press the **Break** button to abort a listing.

2. You can list a subset of variables. You can be specific, as in `list make mpg price`. You can use shorthand: `list m*` means list all variables starting with `m`. `list price-weight` means list variables `price` through `weight` in dataset order.

3. You can abbreviate `list` as `l` (the letter ℓ).

list with in

```
┌─────────────────────────────────────────────────────────────────────────┐
│ □ Stata Results                                                       [X] │
├─────────────────────────────────────────────────────────────────────────┤
│ . list                                                                    │
│                                                                           │
│                     make      price      mpg    weight    gratio   source │
│        1.       VW Rabbit      4697       25      1930      3.78  foreign  │
│        2.          Olds 98     8814       21      4060      2.41  domestic │
│        3.     Chev. Monza      3667        .      2750      2.73  domestic │
│        4.     AMC Concord      4099       22      2930      3.58  domestic │
│        5.      Datsun 510      5079       24      2280      3.54  foreign  │
│        6.                      5189       20      3280      2.93  domestic │
│        7.      Datsun 810      8129       21      2750      3.55  foreign  │
│                                                                           │
│ . l in 1                                                                  │
│                                                                           │
│                     make      price      mpg    weight    gratio   source │
│        1.       VW Rabbit      4697       25      1930      3.78  foreign  │
│                                                                           │
│ . list in -1                                                              │
│                                                                           │
│                     make      price      mpg    weight    gratio   source │
│        7.      Datsun 810      8129       21      2750      3.55  foreign  │
│                                                                           │
│ . list in 2/4                                                             │
│                                                                           │
│                     make      price      mpg    weight    gratio   source │
│        2.          Olds 98     8814       21      4060      2.41  domestic │
│        3.     Chev. Monza      3667        .      2750      2.73  domestic │
│        4.     AMC Concord      4099       22      2930      3.58  domestic │
│                                                                           │
│ . list make mpg in -3/-2                                                  │
│                                                                           │
│                     make       mpg                                        │
│        5.      Datsun 510       24                                        │
│        6.                       20                                        │
│                                                                           │
└─────────────────────────────────────────────────────────────────────────┘
```

Notes

1. in restricts the list to a range of observations.

2. Positive numbers count from the top of the data. Negative numbers count from the end of the data.

3. You may specify both a variable range and an observation range.

list with if

```
┌─┬───────────────────────────────────────────────────────────────────────┬─┐
│─│ Stata Results                                                         │✕│
├─┴───────────────────────────────────────────────────────────────────────┴─┤
│                                                                            │
│ . list                                                                     │
│                                                                            │
│                       make      price      mpg     weight     gratio     source │
│           1.      VW Rabbit      4697       25       1930       3.78    foreign │
│           2.        Olds 98      8814       21       4060       2.41   domestic │
│           3.    Chev. Monza      3667        .       2750       2.73   domestic │
│           4.    AMC Concord      4099       22       2930       3.58   domestic │
│           5.     Datsun 510      5079       24       2280       3.54    foreign │
│           6.                     5189       20       3280       2.93   domestic │
│           7.     Datsun 810      8129       21       2750       3.55    foreign │
│                                                                            │
│ . l if mpg>22                                                              │
│                                                                            │
│                       make      price      mpg     weight     gratio     source │
│           1.      VW Rabbit      4697       25       1930       3.78    foreign │
│           3.    Chev. Monza      3667        .       2750       2.73   domestic │
│           5.     Datsun 510      5079       24       2280       3.54    foreign │
│                                                                            │
│ . l if mpg>22 & mpg~=.                                                      │
│                                                                            │
│                       make      price      mpg     weight     gratio     source │
│           1.      VW Rabbit      4697       25       1930       3.78    foreign │
│           5.     Datsun 510      5079       24       2280       3.54    foreign │
│                                                                            │
│ . l make mpg if mpg>22 | (price>8000 & gratio>3.5)                         │
│                                                                            │
│                       make        mpg                                      │
│           1.      VW Rabbit        25                                       │
│           3.    Chev. Monza         .                                       │
│           5.     Datsun 510        24                                       │
│           7.     Datsun 810        21                                       │
│                                                                            │
│ . l make mpg if mpg>22 | (price>8000 & gratio>3.5) in 1/4                  │
│                                                                            │
│                       make        mpg                                      │
│           1.      VW Rabbit        25                                       │
│           3.    Chev. Monza         .                                       │
│                                                                            │
└────────────────────────────────────────────────────────────────────────────┘
```

Notes

1. `if` expressions may be arbitrarily complicated. `&` means "and"; `|` means "or".

2. `if` may be combined with `in` (or `in` with `if`; the order does not matter).

list with if, continued

3. Missing values are stored internally in Stata as $+\infty$, so `mpg>22` evaluates to true when `mpg` equals missing value. You can exclude observations with missing values for `mpg` by adding '`& mpg~=.`' to your `if` expression. The '`~=`' is the symbol for "not equal", and '`.`' is the missing-value symbol.

4. The logical operators are

<	less than
<=	less than or equal
==	equal
>=	greater than or equal
>	greater than
~=	not equal (!= can also be used)
&	and
|	or
~	not (logical negation; ! can also be used)
()	parentheses specify order of evaluation

By default, `&` is evaluated before `|`; therefore, $a\,|\,b\,\&\,c$ means $a\,|\,(b\,\&\,c)$, which is true if a is true or if b and c are both true. To specify a or b being true, and c being true too, type $(a\,|\,b)\,\&\,c$.

list with if, common mistakes

```
┌─────────────────────────────────────────────────────────────────────────┐
│ ▭ Stata Results                                                       ☒ │
├─────────────────────────────────────────────────────────────────────────┤
│ . list                                                                    │
│                                                                           │
│                     make      price       mpg     weight     gratio       source │
│       1.       VW Rabbit       4697        25       1930       3.78      foreign │
│       2.         Olds 98       8814        21       4060       2.41     domestic │
│       3.     Chev. Monza       3667         .       2750       2.73     domestic │
│       4.     AMC Concord       4099        22       2930       3.58     domestic │
│       5.      Datsun 510       5079        24       2280       3.54      foreign │
│       6.                       5189        20       3280       2.93     domestic │
│       7.      Datsun 810       8129        21       2750       3.55      foreign │
│                                                                           │
│ . list if mpg=21                                                          │
│ =exp not allowed                                                          │
│ r(101);                                                                   │
│                                                                           │
│ . list if mpg==21                                                         │
│                                                                           │
│                     make      price       mpg     weight     gratio       source │
│       2.         Olds 98       8814        21       4060       2.41     domestic │
│       7.      Datsun 810       8129        21       2750       3.55      foreign │
│                                                                           │
│ . list if mpg==21 if weight>4000                                          │
│ invalid syntax                                                            │
│ r(198);                                                                   │
│                                                                           │
│ . list if mpg==21 and weight>4000                                         │
│ invalid 'and'                                                             │
│ r(198);                                                                   │
│                                                                           │
│ . list if mpg==21 & weight>4000                                           │
│                                                                           │
│                     make      price       mpg     weight     gratio       source │
│       2.         Olds 98       8814        21       4060       2.41     domestic │
└─────────────────────────────────────────────────────────────────────────┘
```

Notes

1. Tests of equality are specified with double equal signs, not single—if mpg==21, not if mpg=21. Single equal signs, you will learn, are used for assignment.

2. Joint tests are specified with &, not multiple ifs—if mpg==21 & weight>4000, not if mpg==21 if weight>4000.

3. Use & and |, not the words and and or.

list with if, common mistakes, continued

```
┌─┬──────────────────────────────────────────────────────────────────────┬─┐
│☐│ Stata Results                                                        │✕│
├─┴──────────────────────────────────────────────────────────────────────┴─┤
│                                                                           │
│ . list                                                                    │
│                                                                           │
│                    make      price      mpg     weight     gratio    source│
│       1.      VW Rabbit       4697       25       1930       3.78    foreign│
│       2.         Olds 98      8814       21       4060       2.41   domestic│
│       3.    Chev. Monza       3667        .       2750       2.73   domestic│
│       4.    AMC Concord       4099       22       2930       3.58   domestic│
│       5.      Datsun 510      5079       24       2280       3.54    foreign│
│       6.                      5189       20       3280       2.93   domestic│
│       7.      Datsun 810      8129       21       2750       3.55    foreign│
│                                                                           │
│ . list if make==AMC Concord                                               │
│ AMC not found                                                             │
│ r(111);                                                                   │
│                                                                           │
│ . list if make=="AMC Concord"                                             │
│                                                                           │
│                    make      price      mpg     weight     gratio    source│
│       4.    AMC Concord       4099       22       2930       3.58   domestic│
│                                                                           │
└───────────────────────────────────────────────────────────────────────────┘
```

Note

Tests with strings are allowed, but if you specify the contents of the string, the contents are enclosed in double quotes: if make=="AMC Concord".

list with if, common mistakes, continued

```
┌─────────────────────────────────────────────────────────────────────────┐
│ ▣  Stata Results                                                      ⊠  │
├─────────────────────────────────────────────────────────────────────────┤
│                                                                          │
│  . list if source==domestic                                             │
│  domestic not found                                                     │
│  r(111);                                                                │
│                                                                          │
│  . list if source==0                                                    │
│                                                                          │
│                    make        price       mpg     weight    gratio      source   │
│      2.          Olds 98        8814        21       4060      2.41     domestic  │
│      3.       Chev. Monza       3667         .       2750      2.73     domestic  │
│      4.       AMC Concord       4099        22       2930      3.58     domestic  │
│      6.                         5189        20       3280      2.93     domestic  │
│                                                                          │
│  . list if source==0, nolabel                                           │
│                                                                          │
│                    make        price       mpg     weight    gratio      source   │
│      2.          Olds 98        8814        21       4060      2.41          0     │
│      3.       Chev. Monza       3667         .       2750      2.73          0     │
│      4.       AMC Concord       4099        22       2930      3.58          0     │
│      6.                         5189        20       3280      2.93          0     │
│                                                                          │
└─────────────────────────────────────────────────────────────────────────┘
```

Notes

1. Watch out for value labels—they look like strings but are not. Variable `source` takes on values 0 and 1, and we have merely labeled 0 "domestic" and 1 "foreign" (see Chapter 7).

2. To see the underlying numeric values of variables with labeled values, use the `nolabel` option on the `list` or `edit` command (see [R] **list** and [R] **edit**). You can also determine the correspondence between labels and numeric values with the `label list` command (see [R] **label**).

More

When you see a `--more--` at the bottom of the Results window, it means that there is more information to be displayed. This happens, for example, when `listing` a large number of observations.

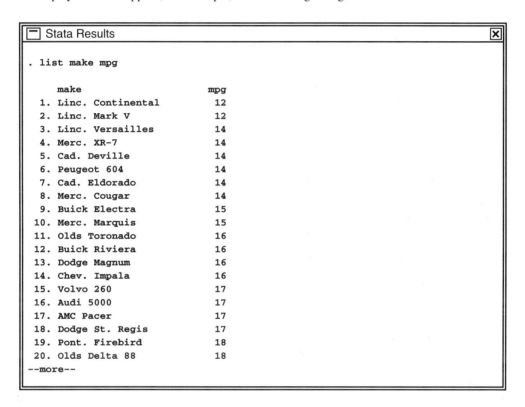

```
 🔲 Stata Results                                                    ☒

 . list make mpg

        make                    mpg
   1. Linc. Continental          12
   2. Linc. Mark V               12
   3. Linc. Versailles           14
   4. Merc. XR-7                  14
   5. Cad. Deville               14
   6. Peugeot 604                14
   7. Cad. Eldorado              14
   8. Merc. Cougar               14
   9. Buick Electra              15
  10. Merc. Marquis              15
  11. Olds Toronado              16
  12. Buick Riviera              16
  13. Dodge Magnum               16
  14. Chev. Impala               16
  15. Volvo 260                  17
  16. Audi 5000                  17
  17. AMC Pacer                  17
  18. Dodge St. Regis            17
  19. Pont. Firebird             18
  20. Olds Delta 88              18
 --more--
```

If you want to see the next screen of text, press any key, such as the *Space Bar*. Or, equivalently, you can click on the **More** button. To see just the next line of text, press *Enter*.

Break

If you want to interrupt a Stata command, click on the **Break** button. Or, equivalently, you can press *Ctrl-Break*.

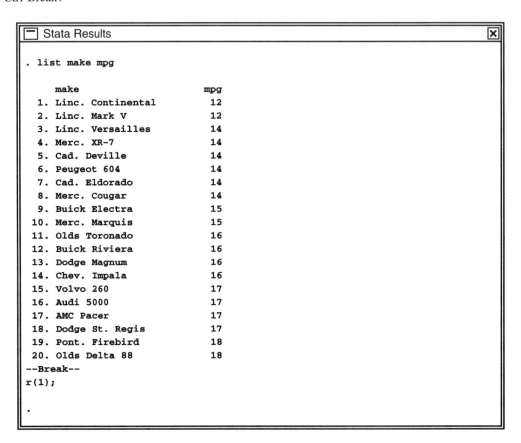

Note

It is always safe to click **Break**. After you click **Break**, the state of the system is the same as if you had never issued the command.

11   Creating new variables

generate and replace

In this chapter, you will learn:

To create a new variable that is an algebraic expression of other variables:	generate *newvar* = *exp*
generate may be abbreviated:	g *newvar* = *exp*
To change the contents of an existing variable:	replace *oldvar* = *exp*
replace may not be abbreviated.	

exp is an algebraic expression that is a combination of existing variables, operators, and functions.

Operators:

	Arithmetic		Logical		Relational (numeric and string)
+	addition	~	not	>	greater than
−	subtraction	|	or	<	less than
*	multiplication	&	and	>=	> or equal
/	division			<=	< or equal
^	power			==	equal
				~=	not equal
+	string concatenation				

Mathematical functions:
   abs(), acos(), asin(), atan(), comb(), cos(), digamma(), exp(), ln(), lnfact(), lngamma(), log10(), mod(), sin(), sqrt(), tan(), trigamma().

Statistical functions:
   Binomial(), binorm(), chiprob(), fprob(), gammap(), ibeta(), invbinomial(), invchi(), invfprob(), invgammap(), invnchi(), invnorm(), invt(), nchi(), normd(), normprob(), npnchi(), tprob(), uniform().

String functions:
   index(), length(), lower(), ltrim(), real(), rtrim(), string(), substr(), trim(), upper().

Date functions:
   date(), day(), dow(), mdy(), month(), year().

Special functions:
   autocode(), cond(), e(sample), float(), group(), int(), max(), min(), missing(), recode(), replay(), round(), sign(), sum().

See [U] **16 Functions and expressions** for a complete list and full details.

generate

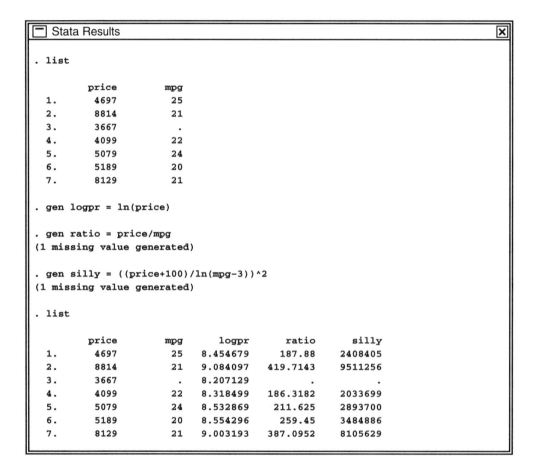

Notes

1. The form of the **generate** command is **generate** *newvar* = *exp*, where *newvar* is a new variable name (it cannot be the name of a variable that already exists) and *exp* is any valid expression.

2. The **generate** command may be abbreviated as **g**, **ge**, **gen**, etc.

3. An expression is a combination of existing variables, operators, and functions. Expressions can be made as complicated as you want.

4. Calculation on a missing value yields a missing value. So does division by zero, etc.

5. If missing values are generated, the number of missing values in *newvar* is always reported. The lack of a mention means no missing values resulted.

replace

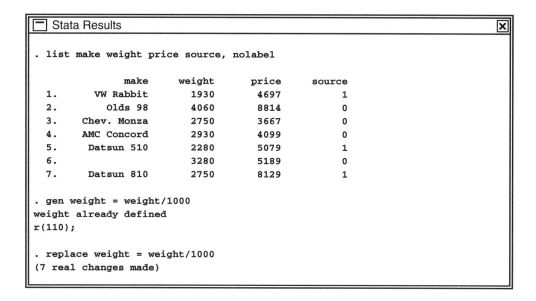

```
. list make weight price source, nolabel

            make    weight     price    source
1.      VW Rabbit      1930      4697         1
2.         Olds 98     4060      8814         0
3.    Chev. Monza      2750      3667         0
4.    AMC Concord      2930      4099         0
5.     Datsun 510      2280      5079         1
6.                     3280      5189         0
7.     Datsun 810      2750      8129         1

. gen weight = weight/1000
weight already defined
r(110);

. replace weight = weight/1000
(7 real changes made)
```

Notes

1. Use `generate` to create new variables and `replace` to change the contents of existing variables. Stata requires this so you do not accidentally modify your data.

2. The `replace` command cannot be abbreviated. Stata generally requires you to spell out completely any command that can alter your existing data.

replace, continued

You want to create a new variable `predpric`, which will be the predicted price of the cars in the following year. You estimate that domestic cars will increase in price by 5% and foreign cars by 10%. First use `generate` to compute the predicted domestic car prices. Then use `replace` to change the missing values for the foreign cars to their proper values.

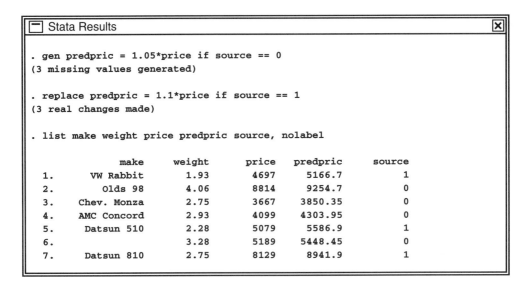

```
┌─ Stata Results ──────────────────────────────────────────────────────────[X]┐
│                                                                              │
│  . gen predpric = 1.05*price if source == 0                                  │
│  (3 missing values generated)                                                │
│                                                                              │
│  . replace predpric = 1.1*price if source == 1                               │
│  (3 real changes made)                                                       │
│                                                                              │
│  . list make weight price predpric source, nolabel                          │
│                                                                              │
│                    make    weight    price   predpric    source              │
│         1.     VW Rabbit      1.93     4697     5166.7         1              │
│         2.       Olds 98      4.06     8814     9254.7         0              │
│         3.   Chev. Monza      2.75     3667    3850.35         0              │
│         4.   AMC Concord      2.93     4099    4303.95         0              │
│         5.    Datsun 510      2.28     5079     5586.9         1              │
│         6.                    3.28     5189    5448.45         0              │
│         7.    Datsun 810      2.75     8129     8941.9         1              │
│                                                                              │
└──────────────────────────────────────────────────────────────────────────────┘
```

generate with string variables

```
┌─────────────────────────────────────────────────────────────────────────┐
│ ⊟  Stata Results                                                     ☒  │
├─────────────────────────────────────────────────────────────────────────┤
│                                                                           │
│  . list make source, nolabel                                              │
│                                                                           │
│                      make      source                                     │
│       1.         VW Rabbit          1                                     │
│       2.           Olds 98          0                                     │
│       3.       Chev. Monza          0                                     │
│       4.       AMC Concord          0                                     │
│       5.        Datsun 510          1                                     │
│       6.                            0                                     │
│       7.        Datsun 810          1                                     │
│                                                                           │
│  . gen orig = "D" if source == 0                                          │
│  type mismatch                                                            │
│  r(109);                                                                  │
│                                                                           │
│  . gen str1 orig = "D" if source == 0                                     │
│  (3 missing values generated)                                             │
│                                                                           │
│  . replace str1 orig = "F" if source == 1                                 │
│  str1 not allowed                                                         │
│  r(101);                                                                  │
│                                                                           │
│  . replace orig = "F" if source == 1                                      │
│  (3 real changes made)                                                    │
│                                                                           │
└─────────────────────────────────────────────────────────────────────────┘
```

Notes

1. "Type mismatch" error occurred because `generate`, by default, creates numeric variables in which a string result cannot be stored. For a string variable, the variable name must be prefixed with the variable type `str1`, `str2`, . . . , `str80`.

2. An existing variable cannot be prefixed with the variable type since the variable type was already set when the variable was created.

generate with string variables, continued

```
┌─────────────────────────────────────────────────────────────────────────┐
│ ⊟ Stata Results                                                      ⊠ │
├─────────────────────────────────────────────────────────────────────────┤
│                                                                           │
│  . gen str20 makeorig = make + " " + orig                                 │
│                                                                           │
│  . gen str8 word2 = substr(make, index(make," ")+1, .)                    │
│  (1 missing value generated)                                              │
│                                                                           │
│  . list make orig makeorig word2                                          │
│                                                                           │
│                   make      orig          makeorig      word2             │
│      1.      VW Rabbit         F       VW Rabbit F     Rabbit             │
│      2.        Olds 98         D         Olds 98 D         98             │
│      3.    Chev. Monza         D     Chev. Monza D      Monza             │
│      4.    AMC Concord         D     AMC Concord D    Concord             │
│      5.     Datsun 510         F      Datsun 510 F        510             │
│      6.                        D                 D                        │
│      7.     Datsun 810         F      Datsun 810 F        810             │
│                                                                           │
└─────────────────────────────────────────────────────────────────────────┘
```

Notes

1. The operator '+' when applied to string variables will concatenate the strings (i.e., join them together). The expression "this" + "that" results in the string "thisthat". When the variable makeorig was generated, a space (" ") was added between the two strings.

2. index(s_1, s_2) produces an integer equal to the first position in s_1 in which s_2 is found, or 0 if not found.

   substr(s, c_0, c_1) produces a string equal to the columns c_0 to $c_0 + c_1$ of s. If $c_1 = .$, it gives a string from c_0 to end of string.

   Therefore, substr(s,index(s," ")+1,.) produces s with its first word removed.

12 Deleting variables and observations

clear, drop, and keep

In this chapter, you will learn:

To drop all the data in memory:	`clear`
or:	`drop _all`
To drop a single variable:	`drop` *varname*
Example:	`drop weight`
To drop any number of variables:	`drop` *varname*(*s*)
Example:	`drop make mpg`
To drop *varname$_i$* through *varname$_j$*:	`drop` *varname$_i$*-*varname$_j$*
Example:	`drop make-mpg`
To drop all variables starting with pop:	`drop pop*`
You may combine all the above:	`drop make-mpg weight pop*`
To drop the first observation:	`drop in 1`
the second:	`drop in 2`
the third:	`drop in 3`
the second from last:	`drop in -2`
the last:	`drop in -1`
or:	`drop in l` (i.e., the letter ℓ)
To drop observations 1 through 3:	`drop in 1/3`
5 to 17:	`drop in 5/17`
3 to third from last:	`drop in 3/-3`
To conditionally drop observations:	`drop if` *exp*
Example:	`drop if mpg>20`
You may combine `if` and `in`:	`drop if mpg>20 in 3/-3`
`keep` works like `drop` except you specify the variables or observations to keep:	`keep if mpg>20 in 3/-3`
To make changes permanent, resave the data:	Choose **Save** under the **File** menu.
or, alternatively, you can type:	`save` *filename*, `replace`

clear and drop _all

Both `clear` and `drop _all` eliminate the data from memory:

```
┌─ Stata Results ──────────────────────────────────────────────────── ×
│
│ . clear
│
│ . list
│
└──────────────────────────────────────────────────────────────────────
```

or

```
┌─ Stata Results ──────────────────────────────────────────────────── ×
│
│ . drop _all
│
│ . list
│
└──────────────────────────────────────────────────────────────────────
```

They differ in that

1. `drop _all` drops the data from memory.

2. `clear` drops the data and it drops all value label definitions, scalars and matrices, constraints and equations, programs, and previous estimation results. `clear`, in effect, resets Stata.

drop

```
┌─┬────────────────────────────────────────────────────────────────────┬─┐
│ ─ │ Stata Results                                                      │✕│
├─┴────────────────────────────────────────────────────────────────────┴─┤
│                                                                         │
│  . use afewcars                                                         │
│  (A few 1978 cars)                                                      │
│                                                                         │
│  . list                                                                 │
│                                                                         │
│                make      price      mpg    weight     gratio     source │
│      1.      VW Rabbit    4697       25      1930       3.78     foreign │
│      2.        Olds 98    8814       21      4060       2.41    domestic │
│      3.     Chev. Monza   3667        .      2750       2.73    domestic │
│      4.     AMC Concord   4099       22      2930       3.58    domestic │
│      5.      Datsun 510   5079       24      2280       3.54     foreign │
│      6.                   5189       20      3280       2.93    domestic │
│      7.      Datsun 810   8129       21      2750       3.55     foreign │
│                                                                         │
│  . drop in 1/3                                                          │
│  (3 observations deleted)                                               │
│                                                                         │
│  . list                                                                 │
│                                                                         │
│                make      price      mpg    weight     gratio     source │
│      1.     AMC Concord   4099       22      2930       3.58    domestic │
│      2.      Datsun 510   5079       24      2280       3.54     foreign │
│      3.                   5189       20      3280       2.93    domestic │
│      4.      Datsun 810   8129       21      2750       3.55     foreign │
│                                                                         │
│  . drop if mpg>21                                                       │
│  (2 observations deleted)                                               │
│                                                                         │
│  . list                                                                 │
│                                                                         │
│                make      price      mpg    weight     gratio     source │
│      1.                   5189       20      3280       2.93    domestic │
│      2.      Datsun 810   8129       21      2750       3.55     foreign │
│                                                                         │
│  . drop gratio                                                          │
│  . list                                                                 │
│                                                                         │
│                make      price      mpg    weight     source            │
│      1.                   5189       20      3280    domestic           │
│      2.      Datsun 810   8129       21      2750     foreign           │
│                                                                         │
└─────────────────────────────────────────────────────────────────────────┘
```

drop, continued

```
┌─────────────────────────────────────────────────────────────────┐
│ ⊟ Stata Results                                            ⊠│
├─────────────────────────────────────────────────────────────────┤
│ . drop m* weight                                                │
│ . list                                                          │
│                                                                 │
│            price       source                                   │
│    1.       5189     domestic                                   │
│    2.       8129      foreign                                   │
│                                                                 │
│ . drop _all                                                     │
│ . list                                                          │
│                                                                 │
└─────────────────────────────────────────────────────────────────┘
```

Note

To make changes permanent, you must resave the data. Since we do not in the above example, afewcars.dta remains unchanged.

keep

```
┌─ Stata Results ──────────────────────────────────────────────────[X]─┐
│                                                                       │
│  . use afewcars, clear                                                │
│  (A few 1978 cars)                                                    │
│                                                                       │
│  . list                                                               │
│                                                                       │
│                     make      price      mpg    weight    gratio    source │
│         1.     VW Rabbit       4697       25      1930      3.78   foreign │
│         2.       Olds 98       8814       21      4060      2.41  domestic │
│         3.   Chev. Monza       3667        .      2750      2.73  domestic │
│         4.   AMC Concord       4099       22      2930      3.58  domestic │
│         5.    Datsun 510       5079       24      2280      3.54   foreign │
│         6.                     5189       20      3280      2.93  domestic │
│         7.    Datsun 810       8129       21      2750      3.55   foreign │
│                                                                       │
│  . keep in 4/7                                                        │
│  (3 observations deleted)                                             │
│                                                                       │
│  . list                                                               │
│                                                                       │
│                     make      price      mpg    weight    gratio    source │
│         1.   AMC Concord       4099       22      2930      3.58  domestic │
│         2.    Datsun 510       5079       24      2280      3.54   foreign │
│         3.                     5189       20      3280      2.93  domestic │
│         4.    Datsun 810       8129       21      2750      3.55   foreign │
│                                                                       │
│  . keep if mpg<=21                                                    │
│  (2 observations deleted)                                             │
│                                                                       │
│  . list                                                               │
│                                                                       │
│                     make      price      mpg    weight    gratio    source │
│         1.                     5189       20      3280      2.93  domestic │
│         2.    Datsun 810       8129       21      2750      3.55   foreign │
│                                                                       │
└───────────────────────────────────────────────────────────────────┘
```

Note

To make changes permanent, you must resave the data. Since we do not in the above example, afewcars.dta remains unchanged.

Notes

13 A sample session

Putting it together

In this chapter, you will learn:

Most of Stata's commands share a common syntax, which is

[by *varlist*:] *command* [*varlist*] [if *exp*] [in *range*] [, *options*]

where items enclosed in square brackets are optional.

Example: `list` of Chapter 10:

```
list
list mpg weight
list if mpg>20
list mpg weight if mpg>20
list in 1/10
list mpg weight in 1/10
list mpg weight if mpg>20 in 1/10
```

by repeats the command for each set of values of its *varlist*:

```
summarize mpg
by rep78: summarize mpg
```

The first command reports the mean, etc., of `mpg`.
The second command reports the mean, etc., of `mpg` for each value of `rep78`.

Some commands allow `by`, others do not.

Determine which are which by selecting **Stata command...** from the **Help** menu, entering *commandname*, and examining the syntax diagram in the help file.

See **Help** for `summarize` and discover `by` is allowed.
See **Help** for `lv` and discover `by` is not allowed.

A single comma sets off a command's options from the rest of the command.

```
graph mpg weight, ylabel xlabel
```

is the correct way to specify options `ylabel` and `xlabel`.

```
graph mpg weight, ylabel, xlabel
```

with commas between options is incorrect.

Warning:

The purpose of this sample session is to show you a little of Stata in action.

Please do not   assume that because we do not demonstrate a feature,
Stata cannot do it.

Stata has lots of other commands.

Please do not   assume that because we demonstrate a command,
we demonstrate all aspects of that command.

Stata has lots of options.

In all cases, see the *Reference Manual* and *Graphics Manual* for a complete description of Stata's features. When in doubt consult the comprehensive index found in the back of the *Reference Manual*.

Sample session

In this chapter, we will use `auto.dta` shipped with Stata. If you wish to follow along, you must load this data. Launch Stata and choose **Open** from the **File** menu. Select the `auto.dta` file from the directory in which you installed Stata.

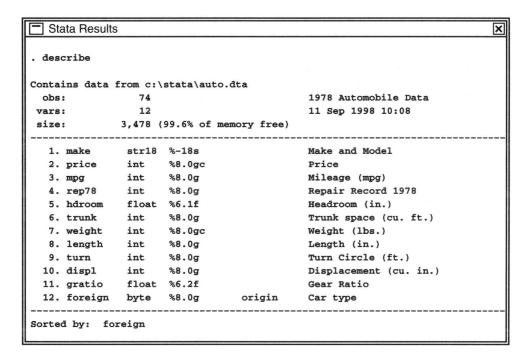

The data that we loaded contains

```
. describe

Contains data from c:\stata\auto.dta
  obs:           74                          1978 Automobile Data
  vars:          12                          11 Sep 1998 10:08
  size:       3,478 (99.6% of memory free)
-------------------------------------------------------------------------
   1. make      str18   %-18s                Make and Model
   2. price     int     %8.0gc               Price
   3. mpg       int     %8.0g                Mileage (mpg)
   4. rep78     int     %8.0g                Repair Record 1978
   5. hdroom    float   %6.1f                Headroom (in.)
   6. trunk     int     %8.0g                Trunk space (cu. ft.)
   7. weight    int     %8.0gc               Weight (lbs.)
   8. length    int     %8.0g                Length (in.)
   9. turn      int     %8.0g                Turn Circle (ft.)
  10. displ     int     %8.0g                Displacement (cu. in.)
  11. gratio    float   %6.2f                Gear Ratio
  12. foreign   byte    %8.0g       origin   Car type
-------------------------------------------------------------------------
Sorted by:  foreign
```

In addition to having more data, the variable we have previously been calling `source` in `afewcars.dta` is now named `foreign`.

Listing can be informative

Here is a little bit of our data:

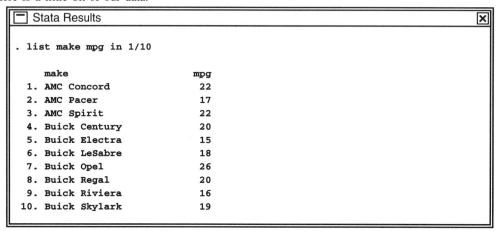

Question: Which cars yield the lowest gas mileage?

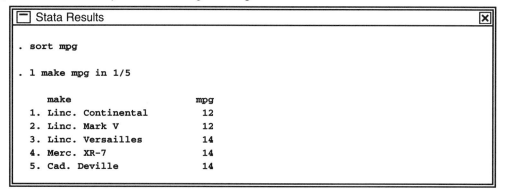

Which 5 cars yield the highest gas mileage?

```
. l make mpg in -5/-1

      make                mpg
70. Toyota Corolla         31
71. Plym. Champ            34
72. Subaru                 35
73. Datsun 210             35
74. VW Diesel              41
```

Descriptive statistics

Question: Not being familiar with 1978 prices, what is the average price of a car in this data?

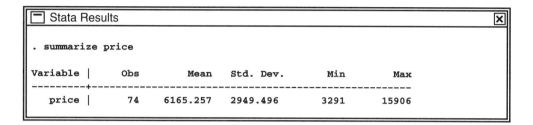

```
┌─────────────────────────────────────────────────────────────────────────┐
│ □ Stata Results                                                       ☒  │
│                                                                           │
│ . summarize price                                                         │
│                                                                           │
│ Variable │      Obs        Mean    Std. Dev.       Min        Max         │
│ ---------+---------------------------------------------------------       │
│    price │       74    6165.257    2949.496       3291      15906         │
│                                                                           │
└─────────────────────────────────────────────────────────────────────────┘
```

Aside: summarize works like list — without arguments it provides a summary of all of the data:

```
┌─────────────────────────────────────────────────────────────────────────┐
│ □ Stata Results                                                       ☒  │
│                                                                           │
│ . summarize                                                               │
│                                                                           │
│ Variable │      Obs        Mean    Std. Dev.       Min        Max         │
│ ---------+---------------------------------------------------------       │
│     make │        0                                                       │
│    price │       74    6165.257    2949.496       3291      15906         │
│      mpg │       74     21.2973    5.785503         12         41         │
│    rep78 │       69    3.405797    .9899323          1          5         │
│   hdroom │       74         3.0         0.8        1.5        5.0         │
│    trunk │       74    13.75676    4.277404          5         23         │
│   weight │       74    3019.459    777.1936       1760       4840         │
│   length │       74    187.9324    22.26634        142        233         │
│     turn │       74    39.64865    4.399354         31         51         │
│    displ │       74    197.2973    91.83722         79        425         │
│   gratio │       74        3.01        0.46       2.19       3.89         │
│  foreign │       74    .2972973    .4601885          0          1         │
│                                                                           │
└─────────────────────────────────────────────────────────────────────────┘
```

Note

make has 0 observations because it is a string — calculating a mean is undefined but not an error. rep78 has only 69 observations because for five cars, it is missing.

Descriptive statistics, continued

Question: What is the average price of cars that are below and above the mean MPG?

```
┌─────────────────────────────────────────────────────────────────────────┐
│ ☐ Stata Results                                                       ☒  │
│                                                                           │
│  . summarize price if mpg<21.3                                            │
│                                                                           │
│  Variable │    Obs       Mean   Std. Dev.       Min       Max            │
│  ---------+-----------------------------------------------------          │
│     price │     43    7091.86   3425.019       3291     15906             │
│                                                                           │
│  . summarize price if mpg>=21.3                                           │
│                                                                           │
│  Variable │    Obs       Mean   Std. Dev.       Min       Max            │
│  ---------+-----------------------------------------------------          │
│     price │     31   4879.968   1344.659       3299      9735             │
│                                                                           │
└─────────────────────────────────────────────────────────────────────────┘
```

Aside: `if` can be suffixed to any command. This is one of Stata's more useful features.

Question: What is the median MPG?

```
┌─────────────────────────────────────────────────────────────────────────┐
│ ☐ Stata Results                                                       ☒  │
│                                                                           │
│  . summarize mpg, detail                                                  │
│                                                                           │
│                        Mileage (mpg)                                      │
│  -------------------------------------------------------------            │
│        Percentiles     Smallest                                           │
│   1%        12            12                                              │
│   5%        14            12                                              │
│  10%        14            14      Obs                   74                │
│  25%        18            14      Sum of Wgt.           74                │
│                                                                           │
│  50%        20                    Mean             21.2973               │
│                        Largest    Std. Dev.        5.785503              │
│  75%        25            34                                              │
│  90%        29            35      Variance         33.47205              │
│  95%        34            35      Skewness        .9487176               │
│  99%        41            41      Kurtosis         3.975005              │
│                                                                           │
└─────────────────────────────────────────────────────────────────────────┘
```

Answer: 20.

Descriptive statistics, continued

Aside: The ', detail' at the end of the `summarize` command is called an option. Most Stata commands share a common syntax:

$$command\ [varlist]\ [\texttt{if}\ exp]\ [\texttt{in}\ range]\ [,\ options\ ]$$

Square brackets mean optional. Thus,

command by itself is valid:	`summarize`
command followed by a *varlist* (variable list) is valid:	`summarize mpg`
	`summarize mpg weight`
command with `if` (with or without a *varlist*) is valid:	`summarize if mpg>20`
	`summarize mpg weight if mpg>20`
And so on.	

If *varlist* is not specified, all the variables are used.

`if` and `in` restrict the data on which the command is run.

options modify what the command does.

Each command's syntax is found in the *Reference Manual* or the *Graphics Manual*. You can learn about summarize in [R] **summarize** or choose **Stata command...** from the **Help** menu and enter `summarize`.

Descriptive statistics, continued

Our data contains variable `foreign` that is 0 if the car was manufactured in the United States or Canada and 1 otherwise.

Problem: Obtain summary statistics for price and MPG for each value of foreign.

There are two solutions to this problem:

1. Type in the commands:

   ```
   summarize price mpg if foreign==0
   summarize price mpg if foreign==1
   ```

2. Or, you could do the following:

```
┌─ Stata Results ──────────────────────────────────────────────────────────[×]─┐
│                                                                               │
│  . sort foreign                                                               │
│                                                                               │
│  . by foreign: summarize price mpg                                            │
│                                                                               │
│  -> foreign=Domestic                                                          │
│  Variable |     Obs        Mean    Std. Dev.       Min        Max             │
│  ---------+-------------------------------------------------------            │
│     price |      52    6072.423    3097.104       3291      15906             │
│       mpg |      52    19.82692    4.743297         12         34             │
│                                                                               │
│  -> foreign= Foreign                                                          │
│  Variable |     Obs        Mean    Std. Dev.       Min        Max             │
│  ---------+-------------------------------------------------------            │
│     price |      22    6384.682    2621.915       3748      12990             │
│       mpg |      22    24.77273    6.611187         14         41             │
│                                                                               │
└───────────────────────────────────────────────────────────────────────────┘
```

Descriptive statistics, continued

Explanation: Stata's general command syntax is

$$\left[\text{by } \textit{varlist}:\right] \textit{command} \left[\textit{varlist}\right] \left[\text{if } \textit{exp}\right] \left[\text{in } \textit{range}\right] \left[, \textit{ options }\right]$$

When by is placed in front of a command, the command is repeated for each set of values of the variables specified.

Some commands allow by and others do not. On-line help always begins with a command's syntax diagram. So, you can quickly tell whether by is allowed by clicking on **Help**, selecting **Stata command...**, and entering *commandname*.

More explanation: Actually, Stata's command syntax includes even more than we have shown you. See [U] **14 Language syntax**.

Note on by: To use by, the data must be sorted by the by-variables. That is why we typed sort foreign before typing by foreign: ....

Minor note: How is it that foreign takes on values 0 and 1 and yet on the output its values appear to be "Foreign" and "Domestic"? foreign has a value label, associating $0 \leftrightarrow$ Domestic and $1 \leftrightarrow$ Foreign; see Chapter 7.

More on by

Problem: It appears that the average MPG of domestic and foreign cars differs. Test the hypothesis that the means are equal.

```
┌─────────────────────────────────────────────────────────────────────────────┐
│ ⊟  Stata Results                                                         ☒  │
├─────────────────────────────────────────────────────────────────────────────┤
│                                                                             │
│  . ttest mpg, by(foreign)                                                   │
│                                                                             │
│  Two-sample t test with equal variances                                     │
│                                                                             │
│  ---------------------------------------------------------------------------│
│     Group |      Obs        Mean    Std. Err.    Std. Dev.   [95% Conf. Interval]│
│  ---------+-----------------------------------------------------------------│
│  Domestic |       52    19.82692     .657777     4.743297    18.50638    21.14747│
│   Foreign |       22    24.77273     1.40951     6.611187    21.84149    27.70396│
│  ---------+-----------------------------------------------------------------│
│  combined |       74     21.2973    .6725511     5.785503     19.9569    22.63769│
│  ---------+-----------------------------------------------------------------│
│      diff |             -4.945804    1.362162                -7.661225   -2.230384│
│  ---------------------------------------------------------------------------│
│  Degrees of freedom: 72                                                     │
│                                                                             │
│                  Ho: mean(Domestic) - mean(Foreign) = diff = 0              │
│                                                                             │
│      Ha: diff < 0                 Ha: diff ~= 0                Ha: diff > 0  │
│        t =  -3.6308                 t =  -3.6308                t =  -3.6308 │
│     P < t =   0.0003            P > |t| =   0.0005           P > t =   0.9997│
│                                                                             │
└─────────────────────────────────────────────────────────────────────────────┘
```

Syntax explanation: The by in this command is not the by prefix described on the previous page; it is a by() option.

by as a prefix—if the command allows it—repeats the command on each subsample of data. If there are three groups, the results will be as if we created three separate datasets and applied the command to each.

by() as an option—if the command allows it—informs the command what groups to use, but the command operates on the data *as a whole*. ttest mpg, by(foreign) tests that the mean of mpg in the groups defined by foreign are equal.

by ...: ttest, were it allowed—and it is not—would perform separate *t* tests within the groups. If ttest allowed this—and one can argue that it should—then by rep78: ttest mpg, by(foreign) would report five equality-of-means tests, one for each value of rep78. Each would be a test of the equality of means for domestic and foreign cars and each would be calculated independently.

More on by, continued

Just remember: `by` in front of a command repeats the command for each group. `by()` on the right of a command does something else, defined in the command's documentation. The key is in the syntax diagrams. The syntax diagram for `summarize` is

$$\left[\texttt{by } varlist:\right] \texttt{ summarize } [varlist] \; [\texttt{if } exp] \; [\texttt{in } range] \; [\texttt{, detail meanonly format}]$$

The `by` is in front—we know what it does.

The syntax diagram for `ttest` is

$$\texttt{ttest } varname \; [\texttt{if } exp] \; [\texttt{in } range], \; \texttt{by}(groupvar) \; [\texttt{unequal welch level}(\#)]$$

No `by` is mentioned in front; therefore it is not allowed. `by()` is mentioned on the right; we must read the documentation to see what it does.

Another aside: We showed only one of the three syntaxes for `ttest`. See the help file for `ttest` or see [R] **ttest** for a complete description of `ttest`.

Analysis note: We have established that in 1978 domestic cars had poorer gas mileage than foreign cars.

Descriptive statistics, making tables

Problem: Obtain counts of the number of domestic and foreign cars.

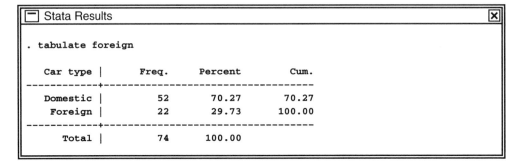

```
. tabulate foreign

  Car type |      Freq.      Percent        Cum.
-----------+-----------------------------------
  Domestic |         52        70.27       70.27
   Foreign |         22        29.73      100.00
-----------+-----------------------------------
     Total |         74       100.00
```

Problem: The data contains variable rep78 recording each car's frequency-of-repair record ($1 =$ poor, ..., $5 =$ excellent). Obtain frequency counts.

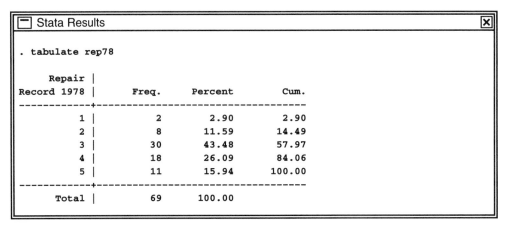

```
. tabulate rep78

    Repair |
Record 1978 |      Freq.      Percent        Cum.
-----------+-----------------------------------
         1 |          2         2.90        2.90
         2 |          8        11.59       14.49
         3 |         30        43.48       57.97
         4 |         18        26.09       84.06
         5 |         11        15.94      100.00
-----------+-----------------------------------
     Total |         69       100.00
```

Problem: We have 74 cars; only 69 have frequency-of-repair records recorded. List the cars for which data is missing.

```
. list make if rep78==.

        make
  3. AMC Spirit
  7. Buick Opel
 45. Plym. Sapporo
 51. Pont. Phoenix
 64. Peugeot 604
```

Descriptive statistics, making tables, continued

Problem: Compare frequency-of-repair records for domestic and foreign cars (i.e., make a two-way table).

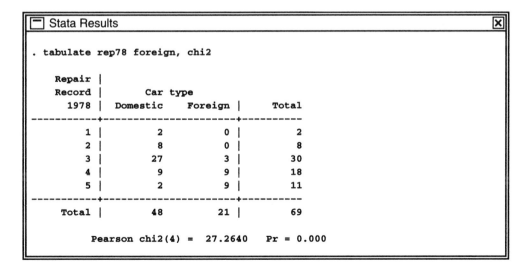

```
  Stata Results                                                   ✕

. tabulate rep78 foreign

    Repair |
    Record |          Car type
      1978 |    Domestic     Foreign |      Total
-----------+----------------------+-----------
         1 |           2           0 |          2
         2 |           8           0 |          8
         3 |          27           3 |         30
         4 |           9           9 |         18
         5 |           2           9 |         11
-----------+----------------------+-----------
     Total |          48          21 |         69
```

Problem: Domestic cars appear to have poorer frequency-of-repair records. Is the difference statistically significant? Obtain a χ^2 (even though there are not at least 5 cars expected in each cell):

```
  Stata Results                                                   ✕

. tabulate rep78 foreign, chi2

    Repair |
    Record |          Car type
      1978 |    Domestic     Foreign |      Total
-----------+----------------------+-----------
         1 |           2           0 |          2
         2 |           8           0 |          8
         3 |          27           3 |         30
         4 |           9           9 |         18
         5 |           2           9 |         11
-----------+----------------------+-----------
     Total |          48          21 |         69

          Pearson chi2(4) =   27.2640   Pr = 0.000
```

Analysis note: We find that frequency-of-repair records differ between domestic and foreign cars. In 1978, domestic cars appear poorer in this regard.

Descriptive statistics, making tables, continued

Aside: `tabulate` provides options to display any or all of the row percentages, column percentages, and cell percentages along with, or instead of, the frequencies.

The option `nofreq` suppresses the frequencies; option `row` adds row percentages; option `col` adds column percentages; and option `cell` adds cell percentages.

```
┌─┐ Stata Results                                                    ⊠
│─│
. tabulate rep78 for, chi2 row col

  Repair |
  Record |        Car type
    1978 |   Domestic     Foreign |      Total
---------+----------------------+----------
       1 |          2           0 |          2
         |     100.00        0.00 |     100.00
         |       4.17        0.00 |       2.90
---------+----------------------+----------
       2 |          8           0 |          8
         |     100.00        0.00 |     100.00
         |      16.67        0.00 |      11.59
---------+----------------------+----------
       3 |         27           3 |         30
         |      90.00       10.00 |     100.00
         |      56.25       14.29 |      43.48
---------+----------------------+----------
       4 |          9           9 |         18
         |      50.00       50.00 |     100.00
         |      18.75       42.86 |      26.09
---------+----------------------+----------
       5 |          2           9 |         11
         |      18.18       81.82 |     100.00
         |       4.17       42.86 |      15.94
---------+----------------------+----------
   Total |         48          21 |         69
         |      69.57       30.43 |     100.00
         |     100.00      100.00 |     100.00

         Pearson chi2(4) =   27.2640   Pr = 0.000
```

Note

We type `tabulate ..., chi2 row col`, not `tabulate ..., chi2, row, col`. One comma sets off the options from the command; commas do not separate options from each other. The order of the options is irrelevant.

Descriptive statistics, correlation matrices

Question: What is the correlation between MPG and weight of car?

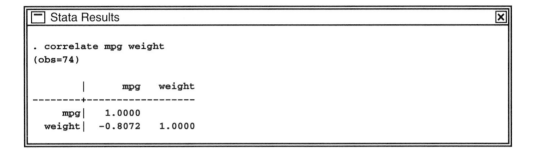

```
. correlate mpg weight
(obs=74)

         |      mpg    weight
---------+------------------
     mpg |   1.0000
  weight |  -0.8072    1.0000
```

Problem: Compare the correlation for domestic and foreign cars.

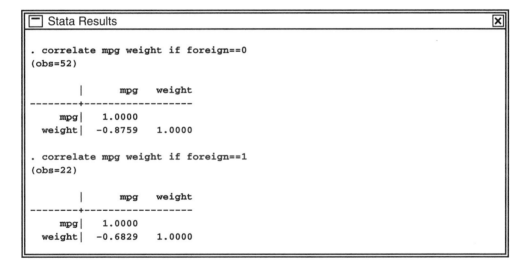

```
. correlate mpg weight if foreign==0
(obs=52)

         |      mpg    weight
---------+------------------
     mpg |   1.0000
  weight |  -0.8759    1.0000

. correlate mpg weight if foreign==1
(obs=22)

         |      mpg    weight
---------+------------------
     mpg |   1.0000
  weight |  -0.6829    1.0000
```

Note

We could have obtained this by typing by foreign: correlate mpg instead.

Descriptive statistics, correlation matrices, continued

Aside: We can produce correlation matrices containing as many variables as we wish.

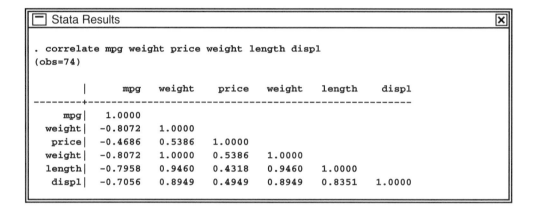

```
Stata Results                                                      [X]

. correlate mpg weight price weight length displ
(obs=74)

         |     mpg   weight    price   weight   length    displ
---------+-----------------------------------------------------
     mpg|  1.0000
  weight| -0.8072   1.0000
   price| -0.4686   0.5386   1.0000
  weight| -0.8072   1.0000   0.5386   1.0000
  length| -0.7958   0.9460   0.4318   0.9460   1.0000
   displ| -0.7056   0.8949   0.4949   0.8949   0.8351   1.0000
```

Graphing data

Problem: We know that the average MPG of domestic and foreign cars differs. We have learned that domestic and foreign cars differ in other ways as well, such as in frequency-of-repair record. We found a negative correlation of MPG and weight—as we would expect—but the correlation appears stronger for domestic cars. Examine, with an eye toward modeling, the relationship between MPG and weight. Begin with a graph.

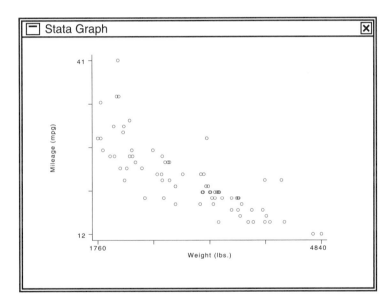

Comment: `graph` is explained in the *Graphics Manual*, but typing `graph` y x draws a graph of y against x. The relationship, we note, is nonlinear.

Note

When you draw a graph, the Graph window appears, probably covering up your Results window. Click on the **Results** button to put your Results window back on top. Want to see the graph again? Click on the **Graph** button. See Chapter 15 for more information about the **Graph** button.

Graphing data, continued

Next, we draw separate graphs for foreign and domestic cars.

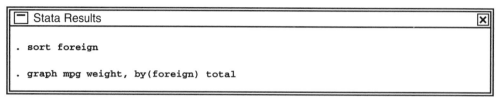

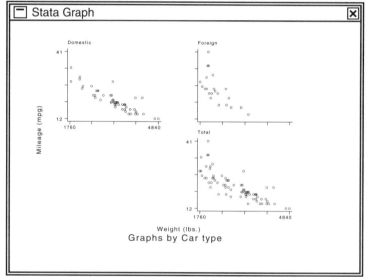

Syntax note: `by()` is on the right of the command, therefore `graph` did whatever it is that it does with the grouping information. What `graph` did is draw separate graphs for domestic and foreign cars in a single image. We have only two groups, but `graph` will allow any number—the individual graphs just get smaller. The `total` option added an overall graph to the image.

If we had placed the `by` in front,

by foreign: graph mpg weight

(`graph` allows this), we would have obtained separate graphs on separate screens for each value of `foreign`.

Analysis note: The relationship is not only nonlinear; the domestic-car relationship appears to differ from that of foreign cars.

Model estimation: linear regression

Restatement of problem: We are to model the relationship between MPG and weight.

Plan of attack: Based on the graphs, we judge the relationship nonlinear and will model MPG as a quadratic in weight. Also based on the graphs, we judge the relationship to be different for domestic and foreign cars. We will include an indicator (dummy) variable for foreign and evaluate afterwards whether this adequately describes the difference. Thus, we will estimate the model:

$$mpg = \beta_0 + \beta_1\, weight + \beta_2\, weight^2 + \beta_3\, foreign + \epsilon$$

foreign is already a 0/1 variable, so we only need create the weight-squared variable:

```
┌─ Stata Results                                                        ☒
│
│  . gen wtsq = weight^2
│
│  . regress mpg weight wtsq foreign
│
│    Source |       SS       df       MS              Number of obs =      74
│  ---------+------------------------------           F(  3,    70) =   52.25
│     Model | 1689.15372        3   563.05124          Prob > F      =  0.0000
│  Residual |  754.30574       70  10.7757963          R-squared     =  0.6913
│  ---------+------------------------------           Adj R-squared =  0.6781
│     Total | 2443.45946       73  33.4720474          Root MSE      =  3.2827
│
│  ------------------------------------------------------------------------------
│       mpg |      Coef.   Std. Err.       t     P>|t|     [95% Conf. Interval]
│  ---------+--------------------------------------------------------------------
│    weight |  -.0165729   .0039692     -4.175   0.000    -.0244892   -.0086567
│      wtsq |   1.59e-06   6.25e-07      2.546   0.013     3.45e-07    2.84e-06
│   foreign |    -2.2035   1.059246     -2.080   0.041      -4.3161   -.0909003
│     _cons |   56.53884   6.197383      9.123   0.000     44.17855    68.89913
│  ------------------------------------------------------------------------------
│
└────────────────────────────────────────────────────────────────────────
```

Model estimation: linear regression, continued

Aside: Stata can estimate many kinds of models, including logistic regression, Cox proportional hazards, etc. Click on **Help**, choose **Search...**, and enter `estimation` for a complete list or look up estimation in the index of the *Stata Reference Manual.*

Continuation of attack: We obtain the predicted values:

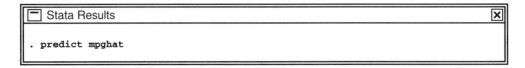

```
. predict mpghat
```

Comment: Be sure to read [U] **23 Estimation and post-estimation commands**. There are a number of features available to you after estimation—one is calculation of predicted values. `predict` just created a new variable called `mpghat` equal to

$$-.0165729\,\texttt{weight} + 1.59 \times 10^{-6}\,\texttt{wtsq} - 2.2035\,\texttt{foreign} + 56.53884$$

Model estimation: linear regression, continued

We can now graph the data and the predicted curve.

Continuation of attack: We just created `mpghat` with `predict`. We could graph the fit and data, but we want to evaluate the fit on the foreign and domestic data separately to determine if our shift parameter is adequate. Thus, we will draw the graphs separately:

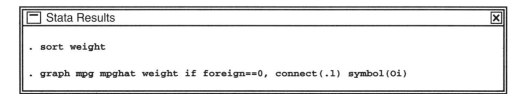

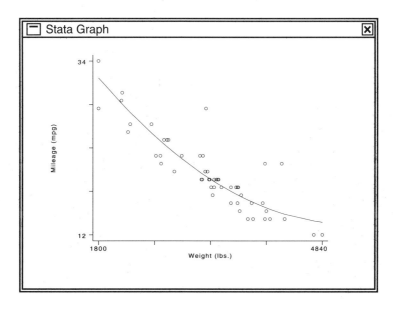

Model estimation: linear regression, continued

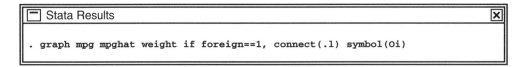

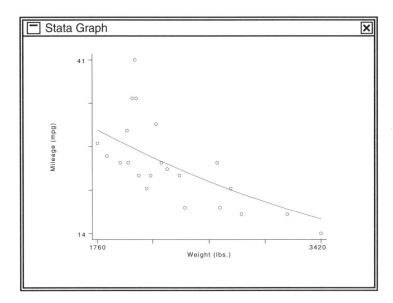

graph mpg mpghat weight says to graph mpg vs. weight and mpghat vs. weight.

connect(.l) says do not connect the mpg vs. weight points—that is the '.'—but do connect (with straight lines) the mpghat vs. weight points—that is the 'l' (ℓ). It is necessary to sort the data by the x-variable—in this case weight—before graphing so that the points are connected in the right order.

symbol(Oi) says use big circles for the mpg vs. weight points—that is the O (capital "oh", not a zero)—but use the invisible symbol (no symbol at all) for the mpghat vs. weight points—that is the 'i'.

Model estimation: linear regression, continued

Problem: You show your results to an engineer. "No," he says. "It should take twice as much energy to move 2,000 pounds 1 mile compared with moving 1,000 pounds, and therefore twice as much gasoline. Miles per gallon is not a quadratic in weight, gallons per mile is a linear function of weight."

You go back to the computer:

```
┌─ Stata Results                                                  ⊠
│
│ . gen gpm = 1/mpg
│
│ . label var gpm "Gallons per mile"
│
│ . sort foreign
│
│ . graph gpm weight, by(foreign) total
│
└
```

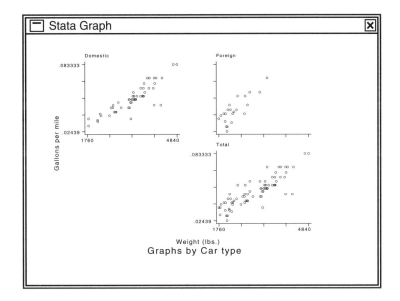

Model estimation: linear regression, continued

Satisfied the engineer is indeed correct, you reestimate:

```
┌─┬─ Stata Results ──────────────────────────────────────────────────────── ⊠ ┐
│                                                                             │
│ . regress gpm weight foreign                                                │
│                                                                             │
│   Source |      SS       df       MS              Number of obs =      74   │
│ ---------+------------------------------          F(  2,    71) =   113.97  │
│    Model | .009117618     2  .004558809           Prob > F      =   0.0000  │
│ Residual | .00284001     71     .00004            R-squared     =   0.7625  │
│ ---------+------------------------------          Adj R-squared =   0.7558  │
│    Total | .011957628    73  .000163803           Root MSE      =   .00632  │
│                                                                             │
│                                                                             │
│ ------------------------------------------------------------------------    │
│      gpm |    Coef.   Std. Err.       t     P>|t|     [95% Conf. Interval]  │
│ ---------+--------------------------------------------------------------    │
│   weight |  .0000163  1.18e-06     13.743   0.000     .0000139    .0000186  │
│  foreign |  .0062205  .0019974      3.114   0.003     .0022379    .0102032  │
│    _cons | -.0007348  .0040199     -0.183   0.855    -.0087504    .0072807  │
│ ------------------------------------------------------------------------    │
│                                                                             │
└─────────────────────────────────────────────────────────────────────────────┘
```

You find foreign cars in 1978 less efficient. Foreign cars may have yielded better gas mileage than domestic cars in 1978, but this was only because they were so light.

14 Using the do-file editor

The do-file editor

In this chapter, you will learn:

Stata for Windows 98/95/NT has a do-file editor.
You can edit do-files and other text files with it.

To enter the do-file editor:
- Click on the **Do-file Editor** button
- Or type `doedit` and press *Enter* in the Command window

The do-file editor lets you submit several commands to Stata at once.
- Type the commands into the editor
- Then press the **Do** button

The do-file editor has standard features found in other text editors.
Cut, **Copy**, **Paste**, **Undo**, **Open**, **Save**, and **Print** are just a few examples.

The do-file editor has more advanced features to aid you in writing do-files.
You will learn about them in this chapter.

The do-file editor toolbar

The do-file editor has eleven buttons. If you ever forget what a button does, hold the mouse pointer over a button for a moment and a box will appear with a description of that button.

New

Start a new do-file.

Open

Open a do-file from disk.

Save

Save to disk the current do-file.

Print

Print the current do-file.

Find

Search for a string in the current do-file.

Continued

The do-file editor toolbar, continued

Cut

Copy the selected text to the clipboard and cut it from the current do-file.

Copy

Copy the selected text to the clipboard.

Paste

Paste text from the clipboard into the current do-file.

Undo

Undo the last change.

Do

Execute the commands in the current do-file.

Run

Execute the commands in the current do-file without showing any output.

Using the do-file editor

Suppose you are about to do an analysis on fuel usage of 1978 automobiles. You know that you will be issuing many commands to Stata during your analysis, and that you want to be able to reproduce your work later without having to type each of the commands again.

You may place commands in a text file; Stata can then read the file and execute each command in sequence. This file is known as a Stata "do-file"; see [U] **19 Do-files**.

> Note: You can create a do-file in any text editor or word processor. You have to be careful to save the file as a text (plain ASCII) file, and you also have to watch out for the text editor placing its own extension on the file. You might tell the text editor to save `myfile.do`, but it may save `myfile.do.txt`. To force such an editor to use the extension you want, enclose the filename in double-quotes; type `"myfile.do"` rather than `myfile.do`.
>
> You can avoid the above hassles by using the do-file editor built-in to Stata for Windows 98/95/NT. Stata's do-file editor knows to use the `.do` extension and will automatically save the file as text (ASCII).

To analyze fuel usage of 1978 automobiles you plan to create a new variable giving gallons per mile. You want to see how that variable changes in relation to vehicle weight for both domestic and imported cars. You decide that graphing the relationship and performing a regression would be good first steps. You want to issue the following commands to Stata:

```
use c:\stata\auto, clear
generate gpm = 1/mpg
label var gpm "Gallons per mile"
sort foreign
graph gpm weight, by(foreign) total
regress gpm weight foreign
```

Using the do-file editor, continued

Click on the **Do-file Editor** button to open the do-file editor. Then type the commands you wish to submit to Stata. Note that we misspelled the variable name on the fourth line—go ahead and intentionally misspell it.

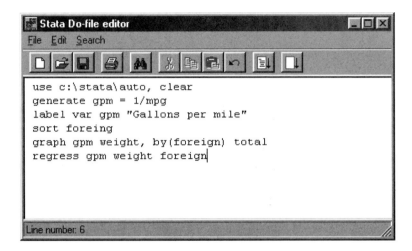

When you are through typing, press the **Do** button:

Using the do-file editor, continued

When you press the **Do** button, Stata executes the commands in sequence and the results appear in the Results window:

```
┌─────────────────────────────────────────────────────────────────────────────┐
│ ☐ Stata Results                                                          ☒  │
├─────────────────────────────────────────────────────────────────────────────┤
│                                                                             │
│ . do "C:\temp\std000001.tmp"                                                │
│                                                                             │
│ . use c:\stata\auto, clear                                                  │
│ (1978 Automobile Data)                                                      │
│                                                                             │
│ . generate gpm = 1/mpg                                                      │
│                                                                             │
│ . label var gpm "Gallons per mile"                                          │
│                                                                             │
│ . sort foreing                                                              │
│ foreing not found                                                           │
│ r(111);                                                                     │
│                                                                             │
│ end of do-file                                                              │
│ r(111);                                                                     │
│                                                                             │
│ .                                                                           │
│                                                                             │
└─────────────────────────────────────────────────────────────────────────────┘
```

The do "C:\... is how Stata executes the commands you typed in the do-file editor. Stata saves the commands to a temporary file and issues the do command to execute them.

Everything worked as planned until Stata saw the misspelled variable. The first three commands were executed, but an error was produced on the fourth. Stata does not know of a variable named foreing. Go back to the do-file editor and correct the mistake.

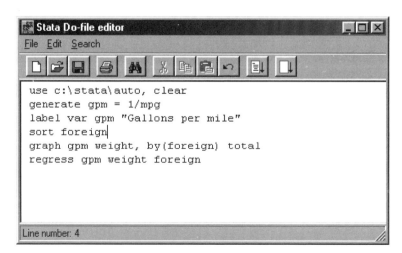

Using the do-file editor, continued

After you correct the spelling of `foreign` and press the **Do** button, Stata will execute each command in the editor, from the beginning.

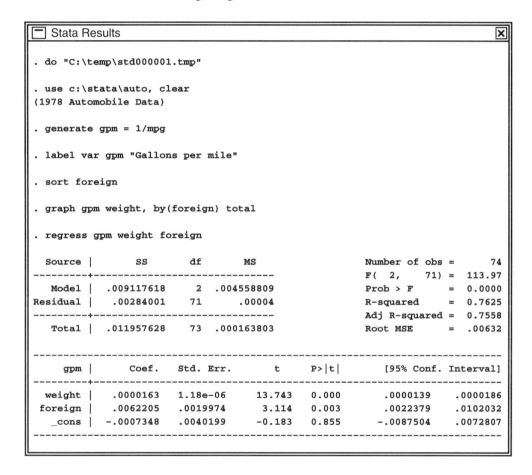

```
⊟ Stata Results                                                                  ☒

. do "C:\temp\std000001.tmp"

. use c:\stata\auto, clear
(1978 Automobile Data)

. generate gpm = 1/mpg

. label var gpm "Gallons per mile"

. sort foreign

. graph gpm weight, by(foreign) total

. regress gpm weight foreign

  Source |       SS       df       MS                  Number of obs =      74
---------+------------------------------               F(  2,    71) =  113.97
   Model |  .009117618     2  .004558809               Prob > F      =  0.0000
Residual |   .00284001    71     .00004                R-squared     =  0.7625
---------+------------------------------               Adj R-squared =  0.7558
   Total |  .011957628    73  .000163803               Root MSE      =  .00632


------------------------------------------------------------------------------
     gpm |      Coef.   Std. Err.       t     P>|t|     [95% Conf. Interval]
---------+--------------------------------------------------------------------
  weight |   .0000163   1.18e-06    13.743    0.000     .0000139    .0000186
 foreign |   .0062205   .0019974     3.114    0.003     .0022379    .0102032
   _cons |  -.0007348   .0040199    -0.183    0.855    -.0087504    .0072807
------------------------------------------------------------------------------
```

A graph was also created, but we do not show that here.

You might want to choose **Save As** from the do-file editor's **File** menu and save this do-file. Later, you could **Open** it (also from the **File** menu) and add more commands as you move forward with your analysis. By saving the commands of your analysis in a do-file as you go, you do not have to worry about retyping them with each new Stata session.

The File menu

The **File** menu of the do-file editor includes standard features found in most text editors. You may choose to start a **New** file, **Open** an existing file, **Save** the current file, or save the current file under a new name with **Save As**. You may also **Print** the current file. There are also buttons on the do-file editor's toolbar which correspond to these features.

There is one other useful feature under the **File** menu: you may choose **Insert File** to insert the contents of a second file at the current cursor position in the do-file editor.

Editing tools

The **Edit** menu of the do-file editor includes the standard **Cut**, **Copy**, and **Paste** capabilities, along with a single-level **Undo**. There are also buttons on the do-file editor's toolbar for easy access to these capabilities. There are several other **Edit** features that you may find useful.

You may delete or select the current line.

You may also shift the current line or selection right or left one tab stop.

The **Change Case** menu item allows you to invert the case of the character to the right of the cursor or to invert an entire selection. When you **Change Case** of an entire selection, the editor will compare the number of lowercase characters with the number of uppercase characters. The case with the greater number of characters will be changed. For example, if you have aBcDe selected, you will have ABCDE after selecting **Change Case**. The editor sees that there are fewer uppercase (2) than lowercase (3) characters, and so makes the entire selection uppercase. If you select **Change Case** again, the entire selection will change to lowercase. There are fewer lowercase (0) than uppercase (5) characters, so all are switched to lowercase.

If there is no selection when you choose **Change Case**, the case of the character immediately to the right of the cursor is switched. The cursor is also moved one character to the right. For example, if the cursor is to the left of 'a' in aBcDe when you select **Change Case**, you will see ABcDe and the cursor will be just to the left of 'B'. If you select **Change Case** again, you will see AbcDe, and the cursor will be just to the left of 'c'.

Preferences

When you select **Preferences** from the **Edit** menu of the do-file editor, you may customize the way the editor behaves.

You may change the font and its size. You may choose any fixed-width TrueType font on your system.

You may set the number of spaces each *Tab* character represents. You may also set whether the do-file editor automatically indents the current line to the same level of indentation as the previous line.

Finally, you may set whether or not the do-file editor automatically saves the current file when you press the **Do** button or the **Run** button. Normally, the file on which you are working is not saved to disk unless you explicitly choose **Save** or **Save As** from the **File** menu. For example, if you **Open** c:\data\myfile.do and make some changes, those changes will not be saved until you save them from the **File** menu. The changes are not saved even when you execute myfile.do by pressing the **Do** button or the **Run** button. This prevents your original file from being overwritten until you tell the do-file editor to overwrite it. If you check **Auto-save on Do/Run** in the **Preferences** dialog, any changes that you have made to myfile.do will be saved when you press either the **Do** button or the **Run** button.

Searching

Stata's do-file editor includes standard **Find** and **Replace** capabilities under the **Search** menu. You should already be familiar with these capabilities from other text editors and word processors. You may access the **Find** capability from a button on the toolbar as well.

The **Search** menu also includes some choices which are useful when you are writing a long do- or ado-file. You may select **Go to Line...** from the **Search** menu and jump straight to any line in your file.

Matching and balancing of parentheses ( ), braces { }, and brackets [ ] are available from the **Search** menu. When you select **Match** from the **Search** menu, the do-file editor looks at the character immediately to the right of the cursor. If it is one of the characters that the editor can match, the editor will find the matching character and place the cursor immediately in front of it. If there is no match, you will hear a beep and the cursor will not move.

When you select **Balance** from the **Search** menu, the do-file editor looks to the left and right of the current cursor position or selection, and creates a selection including the narrowest level of matching characters. If you select **Balance** again, the editor will expand the selection to include the next level of matching characters. If there is no match, you will hear a beep and the cursor will not move.

Balance is easier to explain with an example. Type (now (is the) time) in the do-file editor. Place the cursor between the words is and the. Select **Balance** from the **Search** menu. The do-file editor will select (is the). If you select **Balance** again, the do-file editor will select (now (is the) time).

The Tools menu

You have already learned about the **Do** button. Pressing it is equivalent to choosing **Do** from the **Tools** menu in the do-file editor.

Next to the **Do** button is the **Run** button. Pressing it is equivalent to choosing **Run** from the **Tools** menu. **Run** executes all the commands in the do-file editor just like **Do** does, but **Run** suppresses output. It is unlikely you will ever need to use **Run**; see [U] **19.6.2 Suppressing output** for more details.

Do and **Run** are equivalent to Stata's do and run commands; see [U] **19 Do-files** for a complete discussion.

There are two other useful choices on the **Tools** menu. If you wish to execute a subset of the lines in your do-file, highlight those lines and choose **Do Selection** from the **Tools** menu.

If you wish to execute all commands from the current line through the end of the file, choose **Do to Bottom** from the **Tools** menu.

Saving interactive commands from Stata as a do-file

While working interactively with Stata, you may decide that you would like to rerun the last several commands you typed interactively. You can save the contents of the Review window as a do-file and open that file in the do-file editor. See [GSW] **E.3 Saving the contents of the Review window as a do-file** for details.

Notes

15 Graphs

Working with graphs

In this chapter, you will learn

The **Graph** button brings the Graph window to the top.

To print a graph:
- The Graph window must be open (though it need not be in front)
- Choose **Print Graph** from the **File** menu
- Stata for Windows 98/95/NT users may also click the **Print Graph** button

To save a graph to disk:
- Choose **Save Graph** from the **File** menu
- Enter the *filename*

If you do not specify an extension, .gph will be added and the file named *filename*.gph.

You can save the graph as a Windows Metafile (WMF) by choosing WMF as the file type in the **Save** dialog box.

To load a graph from disk:
- Type graph using *filename*

You can set the font, colors, pen thicknesses, and other options for a graph:
- Pull down the **Prefs** menu, choose **Graph Preferences**, then select an option (see this chapter for details)

Controlling graph properties

There are several options under the menu bar that help you to control the manner in which graphs appear on the screen. Below we detail where these options are located and the purpose that they serve.

Stata uses 10 different pens to draw lines and characters in the graph. You can control both the color and the thickness that those pens use. You can also control the manner in which the graph is sent to the printer. For example, you may desire to use a black background for the graph on the monitor, but not to use black as the background when you print the graph.

If you use Stata for Windows 98/95/NT, you access the graphics options by choosing **Graph Preferences** under the **Prefs** menu. The preferences described below are then accessible from a tabbed dialog box. See [G] **printing Windows** for full details.

If you use Stata for Windows 3.1, you access the graphics options by choosing **Graph** under the **Prefs** menu. The preferences described below are then accessible from individual menu items. See [G] **printing Windows** for full details.

1. **Line Thicknesses**

   You may set the thickness for each of the pens that Stata uses to draw the graph. The default thickness is 1 for each of the pens.

   In general we feel that the printed graphs look better when slightly thicker pens are used. We routinely use thicknesses of 2 and never use any thickness more than 4, but this is in part based on the size that we print graphs (we print them small for this manual). Some experimentation with your printer will allow you to find the optimal settings for your setup. See [G] **pens** for more advice.

2. **Magnifications**

   You may set the magnification of the symbols that Stata uses to draw the graphs. You may also set the magnification for the print size in this dialog.

3. **Colors**

   If you use Stata for Windows 98/95/NT, you will see a tab in the **Graph Preferences** with the color of the background and each pen. You may click on the button next to each color to choose a different color.

   If you use Stata for Windows 3.1, **Colors** leads to menu items of **Pen 1**, **Pen 2**, ... You may change the color for the background and for each pen by selecting the appropriate menu item.

   You can choose colors regardless of whether your printer supports colors since the Windows printer driver will translate all colors except the background to shades of gray. **Print/copy in color** must be checked for Stata to send color information to the printer. See the next page for more details.

4. **Display Thicknesses on Screen**

   This option allows you to toggle whether or not the thicknesses associated with the pens are displayed on the screen. In general the resolution of the screen causes the thicknesses to be displayed much thicker than they appear in printed output. Most users will prefer not to display the thicknesses on the screen.

Controlling graph properties, continued

5. **Print/copy in color**

If this option is not checked, the pen colors all become black in the printed output. We recommend leaving this unchecked for monochrome printers.

6. **Print/copy background color**

If this option is not checked, the background color is ignored in the printed output. This saves on toner if any other color than white is used for your background color. We recommend leaving this unchecked for monochrome printers.

7. **Print Logo**

This option causes the Stata logo to be included at the bottom of the graph when it is printed. Naturally, we recommend this. The logo does not appear on the screen.

8. **Use Windows Fonts**

By default, Stata for Windows uses Windows fonts for the text in graphs. You can change the font style by pulling down the menu box at the top left of the graph window and selecting **Fonts**.

If you uncheck this option, the graph text will be drawn in Stata's own font. It is suggested that you leave this option checked when copying and pasting Stata graphs to other applications; they will be able to understand Windows fonts as text, otherwise they will see Stata's font as a collection of lines and points.

9. **Save Graph Preferences**

This option saves a set of graph preferences under a name of your choice. You do so by selecting **Save Graph Preferences** and clicking on a slot that says ⟨Empty⟩. This opens a dialog where you name the set of preferences. (A name must be 1–8 characters long; any characters including spaces are allowed.) You can later load the named set of preferences using the **Load Graph Preferences** option. A total of 8 sets can be saved. You can change a saved set by overwriting it.

10. **Load Graph Preferences**

This option loads a saved set of graph preferences or restores the default graph settings. Just click on the set of preferences you want.

The Graph button

There is one button that allows you to quickly redisplay graphs that you have already drawn. The **Graph** button brings the graph window to the top.

If you have changed your data and want to redraw the current graph with the new data, simply type `graph` in the Command window; see [G] **redisplaying**.

If you ever decide to close the Stata Graph window, you can only reopen it by reissuing a `graph` command.

Saving and printing graphs

You can save your graphs once they are displayed by choosing **Save Graph** under the **File** menu. In the **Save** dialog box, you can choose to save your graph as either a Stata graph or a Windows Metafile (WMF). You can print the graph by choosing **Print Graph** under the **File** menu. There is also a `gphprint` command which will allow you to automate printing of many graphs. See [G] **printing Windows** for a full explanation.

If you would like to copy your graph to the clipboard in order to import it into another Windows application, you should

1. Display your graph in the Stata Graph window.

2. Click on the title bar of the Stata Graph window.

3. Choose **Copy Graph** from the **Edit** menu.

4. In the other Windows application, you can then choose **Paste** from the **Edit** menu. If this option is not available, then there is most likely another method for importing objects from the clipboard. Consult the documentation for the other application in order to continue.

For more information on printing graphs, see [G] **printing Windows** in the *Graphics Manual*.

16 Logs: Printing and saving output

Putting it on paper and disk

In this chapter, you will learn:

A log is a recording of your Stata session.

Logs are stored as text (ASCII) files.
    You can print them.
    You can load them into a text editor
    or word processor as you would
    any other text file.

To start a log:	Click on the **Open Log** button and then fill in a *filename*. (Windows 3.1 users click on the **Log...** button.)
The log file will be named:	*filename*.`log`
If you specify a file that already exists, you will be asked to either: or:	Append the new log to the file. Overwrite the file with the new log.
The log file is closed automatically when you exit Stata.	
To close the log file sooner:	Click **Close/Suspend Log** and choose **Close log file**. (Windows 3.1 users click **Log...** and choose **Close log file**.)
To temporarily suspend output from being written to the log:	Click **Close/Suspend Log** and choose **Suspend log file**. (Windows 3.1 users click **Log...** and choose **Suspend log file**.)
To resume the suspended log:	Click **Close/Resume Log** and choose **Resume**. (Windows 3.1 users click **Log...** and choose **Resume**.)
To print the current log:	Choose **Print Log** from the **File** menu. Windows 98/95/NT users may also click the **Print Log** button.
To open the Log window: or:	Choose **Log** from the **Window** menu. Click the **Bring Log Window to front** button. (Windows 3.1 users click **Log...** and choose **Bring log window to top**.)
Note: The log *file* is open until you close it even though the Log *window* may be closed.	

Continued

143

Logs

Comments can be added to your log as
    you work. In the Command window,
    type a '*' at the beginning of the line: `* this is a comment`
    This can be useful later: `* report last regression`

Alternatively, you can use the `log` command
    to create a log file.

    To start a new log file, type: `log using` *filename*

    To append to *filename*`.log`, type: `log using` *filename*`, append`

    To overwrite *filename*`.log`, type: `log using` *filename*`, replace`

To close a log file, type: `log close`

By typing: `log using` *filename*`, noproc`
    You can create a log that contains only
    the command lines that you type.
    No output of any kind appears in this log.

A log made with the `noproc` option is useful
    because you can edit it and save it as: *filename*`.do`
    and run it in Stata as a batch file by: Choosing **Do...** from the **File** menu
                                    and entering *filename*

    or by typing: do *filename*

Logging output

All of the output that appears in the Results window can be captured in a log file. The log file is a text (plain ASCII) file that can be printed from Stata or loaded into a text editor or word processor.

1. *To start a log file, you click on the* **Open Log** *button and fill in a name for the file. (Windows 3.1 users click on the* **Log...** *button.)*
   This will open a standard file dialog box allowing you to specify a directory and filename to hold your log. If you do not specify a file extension, the extension .log will be added to the filename.

2. *If you specify a file that already exists:*
   You will be asked if you want to append the new log to the file, or else overwrite the file with the new log.

3. *The log file looks exactly like the output you saw during your Stata session.*

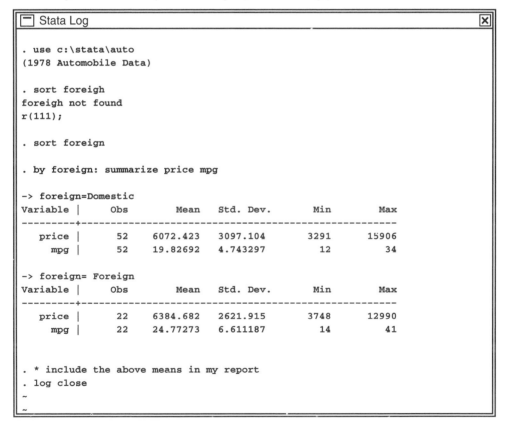

```
┌─ Stata Log ──────────────────────────────────────────────────────── ☒ ┐
│                                                                        │
│  . use c:\stata\auto                                                   │
│  (1978 Automobile Data)                                                │
│                                                                        │
│  . sort foreigh                                                        │
│  foreigh not found                                                     │
│  r(111);                                                               │
│                                                                        │
│  . sort foreign                                                        │
│                                                                        │
│  . by foreign: summarize price mpg                                     │
│                                                                        │
│  -> foreign=Domestic                                                   │
│  Variable |    Obs       Mean    Std. Dev.       Min        Max        │
│  ---------+---------------------------------------------------------   │
│     price |     52   6072.423    3097.104       3291      15906        │
│       mpg |     52   19.82692    4.743297         12         34        │
│                                                                        │
│  -> foreign= Foreign                                                   │
│  Variable |    Obs       Mean    Std. Dev.       Min        Max        │
│  ---------+---------------------------------------------------------   │
│     price |     22   6384.682    2621.915       3748      12990        │
│       mpg |     22   24.77273    6.611187         14         41        │
│                                                                        │
│                                                                        │
│  . * include the above means in my report                             │
│  . log close                                                           │
│  ~                                                                     │
│  ~                                                                     │
└────────────────────────────────────────────────────────────────────── ┘
```

4. *You can scroll through the Log window to view previous output.*

Logging output, continued

5. *To open the Log window:* Choose **Log** from the **Window** menu, or click the **Bring Log Window to front** button if you use Stata for Windows 98/95/NT, or click **Log...** and choose **Bring log window to top** if you use Stata for Windows 3.1. Note: The log *file* is open until you close it even though the Log *window* may be closed.

6. *You can copy from the Log window to the clipboard and paste into the Command window:* Highlight the text you want, pull down the **Edit** menu, and choose **Copy**. To paste into the Command window, choose **Edit—Paste**. Note: If the clipboard contains more than one line, only the first line is pasted into the Command window.

7. *You can also copy from the Log window to the clipboard and paste into the do-file editor (Windows 98/95/NT only):* Highlight the text you want, pull down the **Edit** menu, and choose **Copy**. To paste into the do-file editor, choose **Edit—Paste** from the menu bar of the do-file editor.

8. *You can add comments to your log during your Stata session.*
   If you type a '*' at the beginning of a command line, then the line is treated as a comment. Comments can be very helpful when you later read your log file and try to understand what you did (and why you did it).

```
┌──┬────────────────────────────────────────────────────┬──┐
│ ▬│ Stata Log                                          │ ⊠│
├──┴────────────────────────────────────────────────────┴──┤
│                                                          │
│ . regress mpg weight wtsq gratio displ                   │
│   .                                                      │
│   . (output omitted)                                     │
│   .                                                      │
│ . regress mpg weight wtsq displ                          │
│   .                                                      │
│   . (output omitted)                                     │
│   .                                                      │
│ . regress mpg weight wtsq                                │
│   .                                                      │
│   . (output omitted)                                     │
│   .                                                      │
│ . * the above is the regression I want to report         │
│                                                          │
└──────────────────────────────────────────────────────────┘
```

9. *The log file is closed automatically when you exit Stata.*

10. *Or you can close it whenever you want:*
    Click on the **Close/Suspend Log** button and choose **Close log file**. (Windows 3.1 users click on the **Log...** button and choose **Close log file**.)

11. *For more information about logs, see* [U] **18 Printing and preserving output** *and* [R] **log**.

Printing logs

1. *To print an open log file during a Stata session:*
   Pull down the **File** menu and choose **Print Log**. Stata for Windows 98/95/NT users may also click on the **Print Log** button. After you click **OK**, an **Output Settings** dialog box will appear.

   You can fill in none, any, or all of the items "Header", "Name", and "Project". These items are saved and will appear again in the **Output Settings** dialog (in this and future Stata sessions).

   You can set the font that the printer will use by clicking on **Fonts...** in the **Output Settings** dialog box. The font dialog box will only list the fixed-width "typewriter" fonts (e.g., Courier) available for your printer. Stata, by default, will choose a font size that it thinks is appropriate for your printer.

2. *To print a closed log file:*
   Note: we mean a closed log *file*, not a closed Log *window*. You can print an open log file even though the Log window is closed. To print a closed log file, you must open it first. Click the **Open Log** button (the **Log...** button for Stata for Windows 3.1 users), enter the filename, click **OK**, and then choose **Append** in the next dialog box. Then print the log file according to the above instructions. If you do not want to append to the file, close the log file immediately after printing.

3. *Since the log file is a text file, you can load it into a text editor (like Notepad) or the do-file editor in Stata for Windows 98/95/NT or your favorite word processor.*

4. *From within your editor or word processor, you can edit it—add headings, comments, etc.—format it, and print it.*

5. *Note about word processors:*
   Your word processor will print out the file in its default font. The log file will not be very readable when printed in a proportionally spaced font (e.g., Times Roman or Helvetica). It will look much better printed in a fixed-width "typewriter" font (e.g., Courier).

6. *You may wish to associate the* `.log` *extension with a text editor (like Notepad, Write, or WordPad) in Windows.* You can then edit and print the logs from those Windows applications if you like.

   In Windows 98, 95, or NT, you associate an extension with an application by clicking once on a file with the appropriate extension and then choosing **Options** from the **View** menu. In the **Options** dialog box, click on the **File Types** tab. See the documentation that came with your operating system for more information.

   In Windows 3.1, you associate an extension with an application by choosing **Associate** under the **File** menu of the Windows File Manager. This will open a list box where you may choose an application to associate with the extension. See the documentation that came with Windows 3.1 for more information.

   Once you have done this, anytime you double-click on a file with that extension, the application associated with the extension will be opened and the file loaded into the application.

Rerunning commands as do-files

1. *To create a log file that contains only the command lines that you enter in a Stata session, type*

   ```
   log using myfile, noproc
   ```

2. *No output of any kind—no error messages, etc.—appears in a log file created with the* `noproc` *option.* Here's what the log shown on page 145 would look like if it had been created with the `noproc` option:

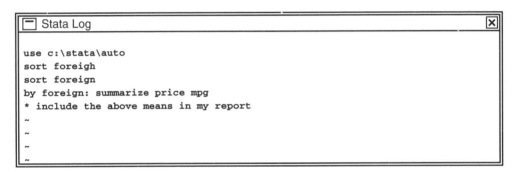

```
use c:\stata\auto
sort foreigh
sort foreign
by foreign: summarize price mpg
* include the above means in my report
~
~
~
~
```

3. *You can also save the contents of the Review window as a do-file.* The Review window stores the last 100 commands you typed. You may find this a more convenient way to create a text file containing only the commands you typed during your session. See [GSW] **E.3 Saving the contents of the Review window as a do-file** for more details.

4. *You can edit a log file created with the* `noproc` *option and rerun the commands in a batch mode from within Stata.* Such files are called do-files, since the batch-mode execution is done with Stata's `do` command. Here's how to create and run a do-file:

   a. Edit the file if necessary. In the previous example, we would want to delete the mistyped command `sort foreigh`. Also, when the log file is closed, the command `log close` is entered into the log as its last line. This command should be deleted; otherwise, running this file as a do-file would automatically close any log you had open. You can open the log file in the do-file editor in Stata for Windows 98/95/NT or in your favorite text editor.

   b. If you have any long command lines—ones that wrap around to two or more lines—you will have to make a few simple changes to your file. See [U] **19.1.3 Long lines in do-files** for details. If none of your command lines extend past one line, you needn't be concerned about this.

Rerunning commands as do-files, continued

c. If you are using a word processor to edit the file, be sure to save the file as a text (plain ASCII) file. You should (but don't have to) rename the file as *filename*.do.

d. To rerun the commands, enter Stata and type do *filename* (or do *filename*.log if you did not change the file's extension to .do). Alternatively, you could choose **Do...** from the **File** menu and then select the file that you want to execute in the dialog.

Note: When you run a do-file, the output will whiz past you without stopping for a --more--, so if you want to see it, you should either have a new log open or insert some more commands in your do-file (see [U] **10 –more– conditions**).

e. See [U] **19 Do-files** and [U] **18 Printing and preserving output** for more information.

Notes

17 Setting font and window preferences

Changing and saving fonts and positions of your windows

In this chapter, you will learn:

You can change fonts in the following windows:

Results	(fixed-width fonts only)
Data Editor	(fixed-width fonts only)
Do-file Editor	(fixed-width TrueType fonts only)
Help and Search	(fixed-width fonts only)
Graph	(TrueType fonts only)
Review	(any font)
Variables	(any font)

(Fixed-width fonts are "typewriter" fonts like Courier.)

To change the font for a window:
- Click once on the window's control-menu box (the little box in the upper-left corner of the window)
- Choose **Font** and a standard Windows font dialog appears
- Select the font
- In the do-file editor, choose **Preferences** from the **Edit** menu and select the font and size

You can resize and rearrange any of Stata's windows.

To set the preferences for viewing and printing graphs:
- See Chapter 15

To save your font, graph, and window preferences:
- Pull down the **Preferences** menu and choose **Save Windowing Preferences**

The next time Stata comes up, it will come up according to these preferences.

To restore the factory settings for the fonts, graph settings, and window arrangements:
- Pull down the **Preferences** menu and choose **Default Windowing**

To go from the factory settings back to the saved preferences:
- Pull down the **Preferences** menu and choose **Load Windowing Preferences**

Note: Only one set of preferences can be saved.

Continued

Closing and opening windows

You can close any window except the Command and Results windows.

If you want to open a closed window or bring a hidden one to the top:
- Pull down the **Window** menu and select the desired window

There are also buttons to bring the Log, Graph, Results, Data Editor, and Do-file Editor windows to the top.

18 How to learn more about Stata

Where to go from here

In this chapter, you will learn:

You should now know enough to begin using Stata.

Play with Stata: | It's the best way to learn.

If you make a mistake and cannot figure out exactly what's wrong, look at the command's help file: | Choose **Help**, select **Stata command...**, and enter *commandname*.

Look at the syntax diagram and examples in the help file and compare them with what you typed.

If you want to find the Stata command that produces a particular statistic, try: | Choose **Help**, select **Search...**, and enter *topic*.

Also, try the index of the *Reference Manual*.

Run the on-line tutorials.
To watch the introductory tutorial, type: | `tutorial intro`

There are a total of 11 on-line tutorials.
This chapter contains a listing of them.

After you have run the tutorials that interest you, we recommend that you read the: | *Stata User's Guide*

The *User's Guide* is designed to be read cover-to-cover, and contains much useful information about Stata.

The *Reference Manual* and *Graphics Manual* are not designed to be read cover-to-cover, but to be sampled when needed. For example, if you plan to do logistic regressions, read about the `logistic` command in: | [R] **logistic**

This chapter contains some advice and recommended reading lists for sampling the *User's Guide* and *Reference Manual*.

Look at the Stata web site; it has much useful information including answers to frequently asked questions (FAQs): | http://www.stata.com

Do you want to learn more?
Take a Stata NetCourse™: | NetCourse 101 is an excellent choice to learn about Stata. See the Stata web site for course information and schedules.

On-line tutorials

Here is a complete listing of Stata's on-line tutorials:

An introduction to Stata:	`intro`
Listing of on-line tutorials:	`contents`
Description of the example datasets we provide:	`ourdata`
How to input your own data into Stata:	`yourdata`
How to make graphs:	`graphics`
How to make tables:	`tables`
Estimating regression models:	`regress`
Estimating one-, two- and N-way ANOVA and ANOCOVA models:	`anova`
Estimating maximum-likelihood logit and probit models:	`logit`
Estimating maximum-likelihood survival models:	`survival`
Estimating factor and principal component models:	`factor`

To run the introductory tutorial (`intro`), type

    tutorial intro

To run the other tutorials, type `tutorial` *filename*. For example, if you want to run `graphics`, type

    tutorial graphics

We recommend you run `intro` first. After that, run the tutorials in whatever order you desire.

If additional official tutorials are added in the future, they will be listed in `tutorial contents`.

Keep in mind that you can press **Break** to end a tutorial at any time. See Chapter 10 for information on **Break**.

Suggested reading from the User's Guide and Reference Manual

The *User's Guide* is designed to be read as a book from cover-to-cover in a linear fashion. The *Reference Manual* is designed as a reference—to be sampled when necessary.

Ideally, after reading this *Getting Started* manual, you should read the *User's Guide* from cover to cover. But you probably want to become a Stata expert right away. Here is a suggested reading list of sections from both the *User's Guide* and the *Reference Manual*. After reading and understanding these sections, you will be well on your way to being a Stata expert.

This list covers fundamental features and also points you to some less obvious features that you might otherwise overlook.

Basic elements of Stata

[U]	Chapter 14	Language syntax
[U]	Chapter 15	Data
[U]	Chapter 16	Functions and expressions

Memory

[U]	Chapter 7	Setting the size of memory
[R]	compress	Compress data in memory

Data input

[U]	Chapter 7	Setting the size of memory
[U]	Chapter 24	Commands to input data
[R]	edit	Edit and list data using data editor
[R]	infile	Quick reference for reading data into Stata
[R]	insheet	Read text (ASCII) data created by a spreadsheet
[R]	append	Append datasets
[R]	merge	Merge datasets

Graphics

Stata Graphics Manual

Useful features that you might overlook

[U]	Chapter 32	Using the Internet to keep up to date
[U]	Chapter 19	Do-files
[U]	Chapter 22	Immediate commands
[U]	Chapter 26	Commands for dealing with strings
[U]	Chapter 27	Commands for dealing with dates
[U]	Chapter 28	Commands for dealing with categorical variables
[U]	Chapter 16.5	Accessing coefficients and standard errors
[U]	Chapter 16.6	Accessing results from Stata commands

Suggested reading, continued

Estimation commands

[U] Chapter 29	Overview of model estimation in Stata
[U] Chapter 23	Estimation and post-estimation commands
[U] Chapter 16.5	Accessing coefficients and standard errors

Basic statistics

[R] anova	Analysis of variance and covariance
[R] ci	Confidence intervals for means, proportions, and counts
[R] correlate	Correlations (covariances) of variables or estimators
[R] egen	Extensions to generate
[R] regress	Linear regression
[R] estimates	Post and redisplay estimation results
[R] predict	Obtain predictions, residuals, etc., after estimation
[R] regression diagnostics	Regression diagnostics
[R] test	Test linear hypotheses after model estimation
[R] summarize	Summary statistics
[R] table	Tables of summary statistics
[R] tabulate	One- and two-way tables of frequencies
[R] ttest	Mean comparison tests

Matrices

[U] Chapter 17	Matrix expressions
[U] Chapter 21.5	Scalars and matrices

Programming

[U] Chapter 19	Do-files
[U] Chapter 20	Ado-files
[U] Chapter 21	Programming Stata
[R] ml	Maximum-likelihood estimation

Internet resources

The Stata web site (http://www.stata.com) is a good place to get more information about Stata. Half of the web site is dedicated to user support. You will find answers to frequently asked questions (FAQs), ways to interact with other users, official Stata updates, free additions to Stata ("Cool ado-files"), and other useful information. You can also subscribe to Statalist, a list server devoted to Stata and statistics discussion.

In addition, you will find information on Stata NetCourses™. A NetCourse is an interactive course offered over the Internet and varies in length from a few to eight weeks. Visit the Stata web site for more information.

If you have access to the web, we suggest you take a quick look at the Stata web site now. You can register your copy of Stata online, and request a free subscription to the *Stata News*.

See Chapter 19 for details on accessing official Stata updates on the Stata web site.

Notes

19 Using the Internet

Internet functionality in Stata

Note: Stata for Windows 98/95/NT includes Internet functionality; Stata for Windows 3.1 does not. Stata for Windows 3.1 has the ability to update itself and to install new features, but must do so from disk; see Chapter 20.

In this chapter, you will learn:

To use Stata to contact Stata's web site for the latest StataCorp news:
> Select **News** from the **Help** menu.

To query Stata's web site to see if you have the latest official updates:
> Select **Official Updates** from the **Help** menu and then
> click on `http://www.stata.com`.

To use Stata to visit web sites that have new routines you can download:
> Select **STB and User-written Programs** from the **Help** menu.

To install an STB insert:
> Select **STB and User-written Programs** from the **Help** menu
> Click on `http://www.stata.com`
> Click on `stb`
>> And then to install insert sg89 from STB-44
> Click on `stb44`
> Click on `sg89`
> Click on `click here to install`

To load a dataset a colleague has placed on a web page:
> type: `use http://www.stata.com/man/example.dta`

To copy a dataset over the web from a colleague:
> type: `copy http://www.stata.com/man/example.dta mycopy.dta`

To look at a text file on a web page:
> type: `type http://www.stata.com/man/readme.txt`

To copy a text file over the web:
> type: `copy http://www.stata.com/man/readme.txt readme.txt`

Official Stata

By official Stata, we mean the pieces of Stata that are provided by us and supported by us. The other and equally important pieces are the user-written additions published in the STB, distributed over Statalist, or distributed in other ways.

Stata can fetch both official updates and user-written programs from the Internet. In the first case, you choose **Official Updates** from the **Help** menu. In the second case, you choose **STB and User-written Programs** from the **Help** menu. (There are command ways of doing this, too; see [U] **32 Using the Internet to keep up to date**.)

Let's start with the official updates. There are two parts to official Stata: the Stata executable and Stata's ado-files.

StataCorp releases updates to official Stata every other month. These updates are to add new features and, sometimes, to fix bugs. Typically the updates are to the ado-files because, in fact, most of Stata is written in Stata's ado language. Occasionally we update the executable as well.

When you choose **Official Updates**, Stata tells you the dates of its two official pieces:

```
----------------------------------------------------------------------------
update
----------------------------------------------------------------------------
Stata executable
    folder:              C:\STATA\
    name of file:        wstata.exe
    currently installed: 12 Jan 1999
Ado-file updates
    folder:              C:\stata\ado\updates\
    names of file:       (various)
    currently installed: 12 Jan 1999
Recommendation
    compare these dates with what is available from
        http://www.stata.com
        other location of your choosing
        diskette (such as an STB diskette)
----------------------------------------------------------------------------
```

If you have an STB diskette, click on `diskette`. If you know of a location which has official updates other than the Stata web site, click on `other location of your choosing` and fill in the site or directory name that you wish.

Assuming you are connected to the Internet, you can click on `http://www.stata.com` to compare the dates of your files with the latest updates available on the Stata web site.

Official Stata, continued

When you click on `http://www.stata.com`, Stata compares your two official pieces with the most up-do-date versions available:

```
--------------------------------------------------------------------------
update query
--------------------------------------------------------------------------
(contacting http://www.stata.com/)
Stata executable
    folder:                C:\STATA\
    name of file:          wstata.exe
    currently installed:   12 Jan 1999
    latest available:      12 Jan 1999
Ado-file updates
    folder:                C:\stata\ado\updates\
    names of file:         (various)
    currently installed:   12 Jan 1999
    latest available:      12 Jan 1999
Recommendation
    Do nothing; all files up-to-date.
--------------------------------------------------------------------------
```

In this case, we are up to date.

We might be told that we need to update our ado-files, or update our executable, or update both. If your computer is not connected to the Internet, Stata will tell you so; see Chapter 20 to learn how to install updates if you are not connected to the Internet.

Updating the official ado-files

Much of Stata is implemented as ado-files—text files containing programs in Stata's ado language. Periodically StataCorp releases updates. The **Official Updates** system makes it easy to keep your official ado-files up-to-date.

After selecting **Official Updates** from the **Help** menu and then clicking on `http://www.stata.com`, you might see

```
----------------------------------------------------------------------------
update query
----------------------------------------------------------------------------
(contacting http://www.stata.com/)
Information about other types of updates may appear . . .
Ado-file updates
    folder:             c:\stata\ado\updates\
    names of files:     (various)
    currently installed: 12 Jan 1999
    latest available:    23 Mar 1999
Recommendation
    update ado-files
```

Stata compares the dates of your ado-files with those available from StataCorp. In this case, Stata is recommending that you update your official ado-files.

> If you do not have write permission for `C:\STATA`, you cannot install official updates in this way. You may still download the official updates, but you will need to use the command-line version of `update`; see [U] **32 Using the Internet to keep up to date** for instructions.

Click on `update ado-files`:

```
----------------------------------------------------------------------------
update ado
----------------------------------------------------------------------------
(contacting http://www.stata.com/)
Stata downloads the new ado-files and tells you about them.
There is nothing more to do.
```

You are done. You can skip to **Updating the executable**.

For your information, Stata does not write over your original ado-files, which are installed in the directory `c:\stata\ado\base`. Instead, Stata downloads updates to the directory `c:\stata\ado\updates`. If there were ever a problem with the updates, you could simply remove the files in `c:\stata\ado\updates`.

Updating the executable

The **Official Updates** system also checks your executable against the most current available:

```
------------------------------------------------------------------------------
update query
------------------------------------------------------------------------------
(contacting http://www.stata.com/)
Stata executable
    folder:              C:\STATA\
    name of file:        wstata.exe
    currently installed: 12 Jan 1999
    latest available:    03 Feb 1999
Information about other types of updates may appear . . .
Recommendation
    update executable
```

In this case, Stata recommends that you update your executable.

If you do not have write permission for C:\STATA, you cannot install official updates in this way. You may still download the official updates, but you will need to use the command-line version of update; see [U] **32 Using the Internet to keep up to date** for instructions.

Click on update executable:

```
------------------------------------------------------------------------------
update executable
------------------------------------------------------------------------------
(contacting http://www.stata.com/)
Stata downloads the new executable and gives you further instructions . . .
```

Stata copies the new executable to C:\STATA\wstata.bin; it does not replace your existing executable. That is for you to do next. Exit all instances of Stata that are running and open a DOS window.

```
C:\WINDOWS> cd \STATA
C:\STATA> rename wstata.exe wstata.old
C:\STATA> rename wstata.bin wstata.exe
```

Once you have verified that the new executable works, you can erase wstata.old.

Downloading user-written programs

Downloading user-written programs is easy. Start by selecting **STB and User-written Programs** from the **Help** menu:

```
-----------------------------------------------------------------------------
Installation and maintenance of STB and user-written programs
-----------------------------------------------------------------------------
User-written programs -- STB, statalist, and others -- are available from a
variety of sources.  Use the links below to find, install, and uninstall them.

    Previously installed packages
        List
        Search...
    New packages
        from http://www.stata.com (includes STB and other links)
        from other sites...
        from diskette (I have an STB or package disk)
```

What to do next will be obvious.

For instance, `lfitx2` is a command from STB-44 sg87. Your Stata has no such command:

```
. lfitx2
unrecognized command
r(199);
. help lfitx2
help for lfitx2 not found
try help contents or search lfitx2
```

You might, however, discover that you want `lfitx2` because you used Stata's `search` command and it sounded interesting.

```
. search windmeijer goodness of fit

STB-44  sg87 . . . . Windmeijer's goodness-of-fit test for logistic regression
        (help lfitx2 if installed) . . . . . . . . . . . . . . . . J. Weesie
        7/98
        alternative to lfit
```

Downloading user-written programs, continued

To obtain sg87 from STB-44, click on `http://www.stata.com` after selecting **STB and User-written Programs** from the **Help** menu.

You should then click on `stb` which will present you with a listing of all STBs available from StataCorp. Click on `stb44`:

```
--------------------------------------------------------------------------
http://www.stata.com/stb/stb44/
STB-44 July 1998
--------------------------------------------------------------------------

DIRECTORIES you could -net cd- to:
    ..                  Other STBs
PACKAGES you could -net describe-:
    dm59                Collapsing datasets to frequencies
    sbe19_1             Tests for publication bias in meta-analysis
    sbe24               metan -- an alternative meta-analysis command
    sg85                Moving summaries
    sg86                Continuation-ratio models for ordinal response data
    sg87                Windmeijer´s goodness-of-fit test for logistic regression
    sg88                Estimating generalized ordered logit models
    sg89                Adjusted predictions and probabilities after estimation
    ssa12               Predicted survival curves for the Cox model
--------------------------------------------------------------------------
```

You may click on any insert to obtain more information. Click on `sg87`:

```
--------------------------------------------------------------------------
package sg87 from http://www.stata.com/stb/stb44
--------------------------------------------------------------------------

TITLE
      STB-44 sg87.  Windmeijer´s goodness-of-fit test for logistic regression.
DESCRIPTION/AUTHOR(S)
      STB insert by Jeroen Weesie, Utrecht University, Netherlands.
      Support: weesie@weesie.fsw.ruu.nl
      After installation, see help lfitx2.
INSTALLATION FILES                            (click here to install)
      sg87/lfitx2.ado
      sg87/lfitx2.hlp
ANCILLARY FILES                               (click here to get)
      sg87/lbw.dta
--------------------------------------------------------------------------
```

Click on `click here to install` to download and install the `lfitx2.ado` and `lfitx2.hlp` files from sg87. (If you want the `lbw.dta` file associated with sg87, click on `click here to get`.)

Downloading user-written programs, continued

You now have lfitx2. Select **Stata command** from the **Help** menu and enter lfitx2:

```
--------------------------------------------------------------------
help for lfitx2                                    (STB-44: sg87)
--------------------------------------------------------------------

Goodness-of-fit after logistic
------------------------------

        lfitx2 [, eps(#)]
The rest of the help file is displayed . . .
```

You can install lots of commands.

You can find out what you have installed by selecting **STB and User-written Programs** from the **Help** menu and clicking on List under Previously installed packages.

Try it after first downloading the STB insert:

```
--------------------------------------------------------------------
directory of installed user-written packages
--------------------------------------------------------------------

[1] package sg87 from http://www.stata.com/stb/stb44
       STB-44 sg87.  Windmeijer's goodness-of-fit test for logistic regression.
--------------------------------------------------------------------
```

If you click on the one-line description of the insert, you will see the full description of the package that you installed:

```
--------------------------------------------------------------------
package sg87 from http://www.stata.com/stb/stb44
--------------------------------------------------------------------

TITLE                                      (click here to uninstall)
       STB-44 sg87.  Windmeijer's goodness-of-fit test for logistic regression.
DESCRIPTION/AUTHOR(S)
       STB insert by Jeroen Weesie, Utrecht University, Netherlands.
       Support: weesie@weesie.fsw.ruu.nl
       After installation, see help lfitx2.
INSTALLATION FILES
       l/lfitx2.ado
       l/lfitx2.hlp
INSTALLED ON
       13 Jan 1999
--------------------------------------------------------------------
```

You can uninstall materials by clicking on click here to uninstall when you are looking at the package description.

Try it. This is all easy.

20 Updating Stata

Maintaining the latest version

In this chapter you will learn:

If you are connected to the Internet and are using Stata for Windows 98/95/NT, you have easy access to the latest official updates. Chapter 19 told you all you need to know about how to keep your Stata current. Skip this chapter.

If you are not connected to the Internet, you can install official updates from the latest STB diskette:
>   Insert the diskette, select **Official Updates** from the **Help** menu, and then click on `diskette`.

You can also install user-written STB inserts from the same diskette:
>   Insert the diskette, select **STB and User-written Programs** from the **Help** menu, and then click on `diskette`.

To do either of the above, you need an STB diskette. Obtain it from us or borrow it from a friend.

Updating the official ado-files

Much of Stata is implemented as ado-files—text files containing programs in Stata's ado language. Every other month StataCorp releases updates to the official ado-files via the STB diskette. The **Official Updates** command makes it easy to keep your official ado-files up-to-date.

After inserting the STB diskette, selecting **Official Updates** from the **Help** menu, and then clicking on `diskette`, you might see

```
  ---------------------------------------------------------------------------
  update query
  ---------------------------------------------------------------------------

  (examining A:\)
  Information about other types of updates may appear . . .

  Ado-file updates
      folder:              c:\stata\ado\updates\
      names of files:      (various)
      currently installed: 12 Jan 1999
      latest available:    23 Mar 1999

  Recommendation
      update ado-files
```

Stata compares the dates of your ado-files with those on the STB diskette. In this case, Stata is recommending that you update your official ado-files.

> If you do not have write permission for `C:\STATA`, you cannot install official updates in this way. You may still install the official updates, but you will need to use the command-line version of `update`; see [U] **32 Using the Internet to keep up to date** for instructions.

Click on `update ado-files`:

```
  ---------------------------------------------------------------------------
  update ado
  ---------------------------------------------------------------------------

  (examining A:\)
  Stata installs the new ado-files and tells you about them.
  There is nothing more to do.
```

You are done. You can skip to **Installing user-written programs**.

For your information, Stata does not write over your original ado-files, which are installed in the directory `c:\stata\ado\base`. Instead, Stata installs updates in the directory `c:\stata\ado\updates`. If there were ever a problem with the updates, you could simply remove the files in `c:\stata\ado\updates`.

Installing user-written programs

Installing user-written programs is easy. Start by selecting **STB and User-written Programs** from the **Help** menu:

```
------------------------------------------------------------------------
Installation and maintenance of STB and user-written programs
------------------------------------------------------------------------

User-written programs -- STB, statalist, and others -- are available from a
variety of sources.  Use the links below to find, install, and uninstall them.
      Previously installed packages
            List
            Search...
      New packages
            from http://www.stata.com (includes STB and other links)
            from other sites...
            from diskette (I have an STB or package disk)
```

What to do next will be obvious.

For instance, `lfitx2` is a command from STB-44 sg87. Your Stata has no such command:

```
. lfitx2
unrecognized command
r(199);
. help lfitx2
help for lfitx2 not found
try help contents or search lfitx2
```

You might, however, discover that you want `lfitx2` because you used Stata's `search` command and it sounded interesting.

```
. search windmeijer goodness of fit
STB-44  sg87 . . . . Windmeijer's goodness-of-fit test for logistic regression
        (help lfitx2 if installed) . . . . . . . . . . . . . . . . J. Weesie
        7/98
        alternative to lfit
```

Installing user-written programs, continued

To obtain sg87 from STB-44, insert the STB-44 diskette and click on `diskette` after selecting **STB and User-written Programs** from the **Help** menu.

```
-----------------------------------------------------------------------------
A:\
STB-44 July 1998
-----------------------------------------------------------------------------
PACKAGES you could -net describe-:
    dm59            Collapsing datasets to frequencies
    sbe19_1         Tests for publication bias in meta-analysis
    sbe24           metan -- an alternative meta-analysis command
    sg85            Moving summaries
    sg86            Continuation-ratio models for ordinal response data
    sg87            Windmeijer's goodness-of-fit test for logistic regression
    sg88            Estimating generalized ordered logit models
    sg89            Adjusted predictions and probabilities after estimation
    ssa12           Predicted survival curves for the Cox model
-----------------------------------------------------------------------------
```

You may click on any insert to obtain more information. Click on **sg87**:

```
-----------------------------------------------------------------------------
package sg87 from A:\
-----------------------------------------------------------------------------
TITLE
      STB-44 sg87.  Windmeijer's goodness-of-fit test for logistic regression.
DESCRIPTION/AUTHOR(S)
      STB insert by Jeroen Weesie, Utrecht University, Netherlands.
      Support: weesie@weesie.fsw.ruu.nl
      After installation, see help lfitx2.
INSTALLATION FILES                              (click here to install)
      sg87/lfitx2.ado
      sg87/lfitx2.hlp
ANCILLARY FILES                                 (click here to get)
      sg87/lbw.dta
-----------------------------------------------------------------------------
```

Click on `click here to install` to install the `lfitx2.ado` and `lfitx2.hlp` files from sg87. (If you want the `lbw.dta` file associated with sg87, click on `click here to get`.)

Installing user-written programs, continued

You now have lfitx2. Select **Stata command** from the **Help** menu and enter lfitx2:

```
-------------------------------------------------------------------------
help for lfitx2                                    (STB-44: sg87)
-------------------------------------------------------------------------
Goodness-of-fit after logistic
------------------------------
        lfitx2 [, eps(#)]
The rest of the help file is displayed . . .
```

You can install lots of commands.

You can find out what you have installed by selecting **STB and User-written Programs** from the **Help** menu and clicking on List under Previously installed packages.

Try it after first installing the STB insert:

```
-------------------------------------------------------------------------
directory of installed user-written packages
-------------------------------------------------------------------------
[1] package sg87 from A:\
      STB-44 sg87.  Windmeijer´s goodness-of-fit test for logistic regression.
-------------------------------------------------------------------------
```

If you click on the one-line description of the insert, you will see the full description of the package that you installed:

```
-------------------------------------------------------------------------
package sg87 from A:\
-------------------------------------------------------------------------
TITLE                                    (click here to uninstall)
      STB-44 sg87.  Windmeijer´s goodness-of-fit test for logistic regression.
DESCRIPTION/AUTHOR(S)
      STB insert by Jeroen Weesie, Utrecht University, Netherlands.
      Support: weesie@weesie.fsw.ruu.nl
      After installation, see help lfitx2.
INSTALLATION FILES
      l/lfitx2.ado
      l/lfitx2.hlp
INSTALLED ON
      13 Jan 1999
-------------------------------------------------------------------------
```

You can uninstall materials by clicking on click here to uninstall when you are looking at the package description.

Try it. This is all easy.

Notes

A Starting and stopping Stata for Windows 98/95/NT

Below we assume that Stata is already installed. If you have not yet installed Stata, follow the installation instructions in Chapter 1.

A.1   Starting Stata

To start Stata:

1. Click on **Start**.

2. Click on **Programs**.

3. Click on **Stata**.

4. Select **Intercooled Stata**, **Pseudo-Intercooled Stata**, or **Small Stata** as appropriate; see [U] **4 Flavors of Stata**.

You will see something like this:

(*Continued on next page*)

173

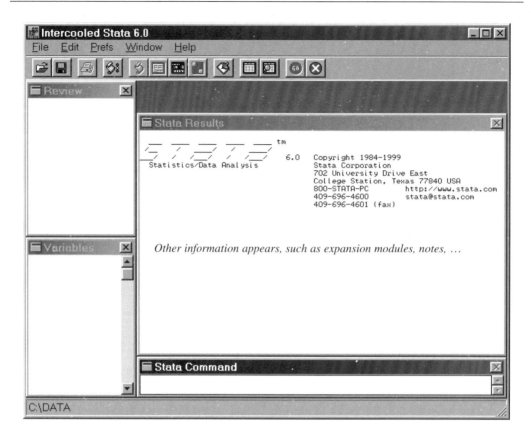

Stata is now waiting for you to type something in the Command window. If Stata does not come up, see [GSW] **C.1 If Stata does not start**.

A.2 Verifying Stata is correctly installed

The first time you start Stata, you should verify that it is installed correctly. Type `verinst` in the Command window and you should see something like

```
. verinst
You are running Intercooled Stata 6.0 for Windows.
Stata is correctly installed.
You can type exit to exit Stata.
```

If you see an error message complaining that the command was not found, then not all of Stata is installed. Go back to Chapter 1 and reinstall Stata.

If you see any other error message, see [GSW] **C.2 verinst problems**.

Remember the `verinst` command. If you ever change your computer setup and are worried that you somehow damaged Stata in the process, you can type `verinst` and obtain the reassuring "Stata is correctly installed" message.

A.3   Exiting Stata

To exit Stata,

1. either click on the close box

   Stata works with a copy of the data in memory. If (1) there is data in memory and if (2) it has changed and if (3) you click to close, a box will appear asking if it is okay to exit without saving the changes.

2. or type `exit` in the Command window.

   If you instead type `exit` and there is changed data in memory, Stata will refuse and instead say, "no; data in memory would be lost". In this case, either you must save the data on disk (see [R] **save**) or you can type `exit, clear` if you do not want to save the changes.

   All of this is designed to prevent you from accidentally losing your data.

As you will discover, the `clear` option is allowed with all potentially destructive commands, of which `exit` is just one example. The command to bring data into memory, `use`, is another example. `use` is destructive since it loads data into memory and, in the process, eliminates any data already there.

If you type a destructive command and the data in memory has been safely stored on disk, Stata performs your request. If your data has changed in some way since it was last saved, Stata responds with the message "no; data in memory would be lost". If you want to go ahead anyway, you can retype the command and add the `clear` option. Once you become familiar with Stata's editing keys, you will discover it is not necessary to physically retype the line. You can press the *PrevLine* key (*PgUp*) to retrieve the last command you typed and append `, clear`.

Of course, you need not wait for Stata to complain—you can add the `clear` option the first time you issue the command—if you do not mind living dangerously.

A.4   The Windows 98/95/NT Properties Sheet

When you click on an icon to start an application in Windows, you are actually executing the instructions in that application's shortcut. The shortcut is defined on the icon's Properties Sheet.

To see the Properties Sheet for any icon, you click once on the icon to highlight it and then pull down **File** and choose **Properties**. Alternatively, you can right-click (click on the right mouse button) on the icon and then choose **Properties**.

The icons for the applications listed under the **Start** button are buried in

   `C:\WINDOWS\Start Menu\Programs`.

Start by clicking on **My Computer** and work your way there. You will find a **Stata** folder. Click on that. You will now see icons for whatever versions of Stata you have installed—probably there is only one.

When you start Stata from the **Start** button, it is exactly the same as clicking on one of these icons.

Now that you have found the **Start** button icons, right-click on one of the Stata icons and choose **Properties**. The icon's Properties Sheet will be displayed. What you will be looking at is that icon's **General** tab and the information there is not very interesting. Click on the **Shortcut** tab. You will see something that contains the following information:

Intercooled Stata

Target type:	Application
Target location:	Stata
Target:	`c:\stata\wstata.exe /k1000`
Start in:	`c:\data\`
Shortcut key:	`None`
Run:	`Normal window`

The names and locations of files may vary from this. For instance, Small Stata users will see a different filename for the **Target** and will not see the **/k1000**.

There are two things to pay attention to: the **Target** and **Start in**. **Target** is the actual command that is executed to invoke Stata. **Start in** is the directory to switch to before invoking the application.

You can change these fields and then click **OK** to save the updated Properties Sheet.

A.5   Starting Stata from other folders

You can have Stata start in whatever directory you desire. Just change the **Start in** field of Stata's Properties Sheet.

Of course, once Stata is running, you can change directories whenever you wish; see [R] **cd**.

A.6   Specifying the amount of memory allocated

The **/k** option in the **Target** field of Stata's Properties Sheet determines the amount of memory allocated to Stata. This memory allocation can be increased or decreased by changing the **/k** option. Only Intercooled and Pseudo-Intercooled Stata users can do this; Small Stata has no **/k** and the memory allocation is fixed.

The **/k1000** in the **Target** field is what tells Stata to allocate 1 megabyte of memory for data.

If you now want Stata to allocate 2 megabytes of memory, change the **/k1000** to **/k2000**.

If you want 32 megabytes, change the field to **/k32000**.

If you want 100 megabytes, change the field to **/k100000**.

Note, if you make the amount too large, when Stata attempts to come up you will see the message "Cannot allocate __K of memory; decrease /k# under **File Properties**".

You can also change the amount of memory once Stata is running by using the `set memory` command; see [GSW] **D. Setting the size of memory**.

A.7   Executing commands every time Stata is started

Stata looks for the file `profile.do` when it is invoked and, if it finds it, executes the commands in it. Stata looks for `profile.do` along your PATH and in the current directory; we recommend that you put `profile.do` in your working directory, `C:\DATA`.

Say that every time you started Stata you wanted `matsize` set to 100 (see [R] **matsize**) and you wanted to `set more off` (see [R] **more**). Create file `C:\DATA\profile.do` containing

```
set matsize 100
set more off
```

When you invoke Stata, these commands will be executed:

(usual opening appears, but with the addition)
`running C:\DATA\profile.do ...`

`. _`

`profile.do` is treated just as any other do-file once it is executed; results are literally as if you started Stata and then typed '`run profile.do`'. The only special thing about `profile.do` is that Stata looks for it and runs it automatically.

See [U] **19 Do-files** for an explanation of do-files. They are nothing more than text (ASCII) files containing a sequence of commands for Stata to execute.

A.8   Making shortcuts

You can arrange to start Stata without going through the **Start** button. On your standard Windows screen, do you see how **My Computer** just sits on the desktop? You can put a Stata icon on the desktop, too.

The process for doing this is to find the Stata executable (not the shortcut we found above) and then drag it with the *right* mouse button where we want it. Here are the details:

1. Open the `c:\stata\` folder or whatever folder you installed Stata into.

2. In the folder, find the executable for which you want a new shortcut. The filenames are

Intercooled Stata:	`wstata.exe`
Pseudo-Intercooled Stata:	`wstatanc.exe`
Small Stata:	`wsmstata.exe`

    Click with the *right* mouse button and drag the appropriate executable onto the desktop.

3. When you release the mouse button, you will see a menu. Choose **Create Shortcut(s) Here**.

You have now created a shortcut. If you want the shortcut in a folder rather than on the desktop, you can drag it into whatever folder appeals to you.

You set the properties for this shortcut just as you would normally. Click with the right mouse button and choose **Properties**. You edit the Properties Sheet as explained in [GSW] **A.4 The Windows 98/95/NT Properties Sheet**.

Note that the Properties Sheet for this new icon is different from the one for Stata under the **Start** button. The new icon could start Stata with 32 megabytes of memory, say, while the **Start** button still uses 1 megabyte.

If you create additional icons, each is different. Some users create separate icons for starting Stata with differing amounts of memory.

A.9   Executing Stata in background (batch) mode

You can run large jobs in Stata in batch mode. To do so, open a DOS window, change to your data directory, and type

```
C:\DATA> c:\stata\wstata /b do bigjob
```

This tells Stata to execute the commands in `bigjob.do`, suppress all screen output, and route the output to `bigjob.log` in the same directory.

If the do-file loads datasets that require more than 1 MB of memory, you will need to allocate this to Stata when you start it. Typing

```
C:\DATA> c:\stata\wstata /k5000 /b do bigjob
```

will run `bigjob.do` with 5 MB of memory. Alternatively, `bigjob.do` can try to increase the memory allocated to Stata while it is running with the `set memory` command; see [GSW] **D. Setting the size of memory**.

While the do-file is executing, the Stata icon will appear on the taskbar together with a rough percentage of how much of `bigjob.do` Stata has executed. (Note that Stata calculates this percentage based on the number of characters in `bigjob.do`, so the percentage may not accurately reflect the amount of time left for the job to complete.)

If you click the icon on the taskbar, Stata will display a box asking if you want to cancel the batch job.

Once the do-file is complete, Stata will flash the icon on the taskbar on and off. You can then click the icon to close Stata.

Note that you do not have to run large do-files in batch mode. Any do-file that you run in batch mode can also be run interactively. Simply start Stata, type `log using` *filename*, and then type `do` *filename*. You can then watch the do-file running, or you can minimize Stata while the do-file is running.

A.10   Launching by double-clicking on a .dta dataset

The first time you start Stata for Windows 98/95/NT, Stata registers with Windows the actions to perform when you double-click on certain types of files. For example, if you double-click on a Stata dataset (`.dta` file), Stata will start and attempt to `use` the file. If you are using Intercooled Stata, the default amount of memory allocated to data will be 1 megabyte (a `/k1000` option is on the command-line for the dataset double-click open action).

To change the amount of memory, you must modify the **Properties** for `.dta` datasets:

1. Double-click on **My Computer**. Pull down **View**. Select **Options**.

2. Click on the **File Types** tab. Scroll through the list until you find **Stata Dataset**. Click once on it, then click the **Edit...** button.

3. You will see an **Edit File Type** dialog. In the Actions section, click on **open** (which is the only thing there), then click the **Edit...** button.

4. In the resulting dialog box, there will be an edit field containing something like

   ```
   c:\stata\wstata.exe /k1000 use "%1"
   ```

   This is the DOS-like command to start Stata with a dataset. Notice its similarities with the **Target** of a regular Properties Sheet. When you double-click on a Stata dataset, Windows executes this command line, replacing the **%1** with the path and filename of the file on which you double-clicked.

   The value of **/k** specifies the amount of memory to allocate. If you want Stata to start with 2 megabytes of memory rather than 1, replace the **/k1000** with **/k2000**.

5. Once you have set **/k** as you wish, click **OK**, then click **Close** from the **Edit File Type** dialog. Then click **Close** from the **Options** dialog.

Note that the **/k** option set here is for all **.dta** datasets, not a specific one. Also note, Small Stata users cannot vary the amount of memory Stata allocates to itself.

What if you want Stata to do other things as well? What if you have written a Stata program called **myuse** (see [U] **20 Ado-files**) which contains a list of commands that you want Stata to execute when it loads a dataset on which you double-click?

The trick is to change the command to be executed from

   ```
   C:\STATA\WSTATA.EXE /k1000 use "%1"
   ```

to

   ```
   C:\STATA\WSTATA.EXE /k1000 myuse "%1"
   ```

myuse.ado might read

—————————————————————————————— top of myuse.ado —————————
```
...
if "`1'"~="" {
        use "`1'"
}
...
```
—————————————————————————————— end of myuse.ado —————————

where ... represents the other commands in **myuse.ado** file. That is, we just add these three lines to the file at the point where we wish to have the dataset loaded.

A.11 Launching by double-clicking on a do-file

Double-clicking on a do-file works just like double-clicking on a dataset. When you double-click on a do-file, Stata comes up in interactive mode, executes the do-file, and then issues a prompt so that you can continue the session or exit.

You set the memory for a do-file in the same manner as you do for a **.dta** dataset. This time when you search for the file type, you click on **Stata do-file** rather than **Stata Dataset**.

The edit field containing the command will say **do** rather than **use**:

   ```
   c:\stata\wstata.exe /k1000 do "%1"
   ```

Note that the /k option set here is for all invocations by double-clicking on a do-file, not a specific one, and it is unrelated to whatever value you may have set for double-clicking on a .dta dataset. Also note that Small Stata users cannot vary the amount of memory Stata allocates to itself.

If you click with the right mouse button on a do-file, a menu appears and you can select **Open** or **Edit**. **Open** does the same as double-clicking on the do-file. **Edit** starts Notepad and loads the do-file. You can edit the action for **Edit** in the same way as described above for the **Open** action if you wish to use an editor other than Notepad.

A.12   Launching by double-clicking on a .gph graph file

When you double-click on a Stata .gph file, Stata comes up and displays the graph. The current directory is the directory containing the graph. Stata remains running. Type `exit` or click to close when you are through with the session.

When you single-click on a Stata .gph file, you can pull down **File** and select **Print** or **Print to default**. If you select **Print**, Stata comes up, displays the graph, and starts a print dialog. If you select **Print to default**, Stata comes up and prints the graph to the default printer, bypassing the print dialog. When the graph is finished printing (or you cancel the print), Stata exits.

Equivalent to single-clicking, pulling down **File**, and choosing **Print** or **Print to default** is to right-click and choose **Print** or **Print to default**.

A.13   Running simultaneous Stata sessions

Each time you double-click on the Stata icon or launch Stata in any other way, you invoke a new instance of Stata, so if you want to run multiple Stata sessions simultaneously, you may.

B Starting and stopping Stata for Windows 3.1

Contents:

Below we assume that Stata is already installed. If you have not yet installed Stata, follow the installation instructions in Chapter 1.

B.1   Starting Stata

To start Stata, double-click on the **Stata** icon. You will see something like this:

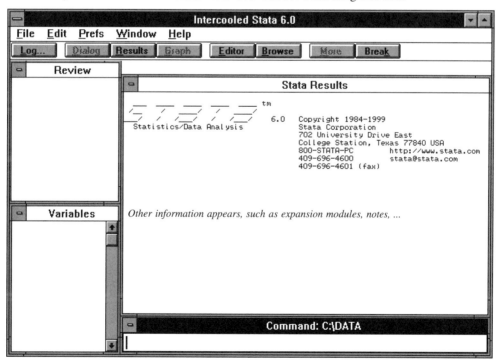

Stata is now waiting for you to type something in the Command window. If Stata does not come up, see [GSW] **C.1 If Stata does not start**.

B.2 Verifying Stata is correctly installed

The first time you start Stata, you should verify that it is installed correctly. Type `verinst` in the Command window and you should see something like

```
. verinst
You are running Intercooled Stata 6.0 for Windows.
Stata is correctly installed.
You can type exit to exit Stata.
```

If you see an error message complaining that the command was not found, then not all of Stata is installed. Go back to Chapter 1 and reinstall Stata.

If you see any other error message, see [GSW] **C.2 verinst problems**.

Remember the `verinst` command. If you ever change your computer setup and are worried that you somehow damaged Stata in the process, you can type `verinst` and obtain the reassuring "Stata is correctly installed" message.

B.3 Exiting Stata

To exit Stata,

1. either double-click on the close box

    Stata works with a copy of the data in memory. If (1) there is data in memory and if (2) it has changed and if (3) you click to close, a box will appear asking if it is okay to exit without saving the changes.

2. or type `exit` in the Command window.

    If you instead type `exit` and there is changed data in memory, Stata will refuse and instead say, "no; data in memory would be lost". In this case, either you must save the data on disk (see [R] **save**) or you can type `exit, clear` if you do not want to save the changes.

All of this is designed to prevent you from accidentally losing your data.

As you will discover, the `clear` option is allowed with all potentially destructive commands, of which `exit` is just one example. The command to bring data into memory, `use`, is another example. `use` is destructive since it loads data into memory and, in the process, eliminates any data already there.

If you type a destructive command and the data in memory has been safely stored on disk, Stata performs your request. If your data has changed in some way since it was last saved, Stata responds with the message "no; data in memory would be lost". If you want to go ahead anyway, you can retype the command and add the `clear` option. Once you become familiar with Stata's editing keys, you will discover it is not necessary to physically retype the line. You can press the *PrevLine* key (*PgUp*) to retrieve the last command you typed and append `, clear`.

Of course, you need not wait for Stata to complain—you can add the `clear` option the first time you issue the command—if you do not mind living dangerously.

B.4   The Windows 3.1 Properties Sheet

When you double-click on an icon to start an application in Windows 3.1, you are actually executing the instructions on the applications shortcut. The shortcut is defined on the application's Properties Sheet.

To see the Properties Sheet for any icon, you click once on the icon to highlight it and then pull down **File** from Program Manager and choose **Properties**. If you do this with **Stata**'s icon, you will see something that contains the following information:

Description: Intercooled Stata
Command Line: c:\stata\wstata /k1000
Working Directory: c:\data
Shortcut Key: none

You may need to click in the individual edit boxes and press the right-arrow key to see all of the information.

The names of the files and directories will vary. For instance, Small Stata users will see a different filename on the command line and there will be no /k1000. In this Intercooled Stata example, we tell Windows to switch to directory c:\data before starting Stata, so that will be the current directory when we are in Stata.

You can change these fields and then click **OK** to save the updated Properties Sheet.

B.5   Starting Stata from other directories

You can have Stata start in whatever directory you desire. Just modify the **Working Directory** field of Stata's Properties Sheet.

Of course, once Stata is running, you can change directories whenever you wish; see [R] **cd**.

B.6   Specifying the amount of memory allocated

The /k in the **Command Line** field of Stata's Properties Sheet determines the amount of memory allocated to Stata. This memory allocation can be increased or decreased by changing the /k option. Only Intercooled and Pseudo-Intercooled Stata users can do this; Small Stata has no /k and the memory allocation is fixed.

The /k1000 in the **Command Line** field is what tells Stata to allocate 1 megabyte of memory to itself when it comes up.

If you want Stata to come up with 2 megabytes of memory, change the /k1000 to /k2000.

If you want 32 megabytes, change the field to /k32000.

If you want 100 megabytes, change the field to /k100000.

Note, if you make the amount too large, when Stata attempts to come up you see the message "Cannot allocate __K of memory; decrease /k# under **File Properties**".

❏ Technical Note

You can set /k to more than the physical amount of memory on your computer if you have virtual memory enabled. Virtual memory is slow but adequate in rare cases when you have a dataset that is too large to load into memory. If you use large datasets frequently, we recommend that you add more memory to your computer.

To determine whether virtual memory is enabled, open the **Main** program group by double-clicking on its icon and then double-click on the **Control Panel** icon. Double-click on the **386 Enhanced** icon. Press the **Virtual Memory** button and then follow instructions. See your Windows documentation for more information.

❏

B.7   Executing commands every time Stata is started

Stata looks for the file `profile.do` when it is invoked and, if it finds it, executes the commands in it. Stata looks for `profile.do` along your PATH and in the current directory; we recommend that you put `profile.do` in your working directory, `C:\DATA`.

Say that every time you started Stata you wanted `matsize` set to 100 (see [R] **matsize**) and you wanted to `set more off` (see [R] **more**). Create file `C:\DATA\profile.do` containing

```
set matsize 100
set more off
```

When you invoke Stata, these commands will be executed:

```
(usual opening appears, but with the addition)
running C:\DATA\profile.do ...

. _
```

`profile.do` is treated just as any other do-file once it is executed; results are literally as if you started Stata and then typed '`run profile.do`'. The only special thing about `profile.do` is that Stata looks for it and runs it automatically.

See [U] **19 Do-files** for an explanation of do-files. They are nothing more than text (ASCII) files containing a sequence of commands for Stata to execute.

B.8   Creating multiple Stata icons

You may create multiple **Stata** icons and have different memory sizes or different working directories associated with each. It is not necessary to create multiple icons if you merely wish to run multiple Stata sessions simultaneously; each time you double-click on a **Stata** icon, you bring up a new instance of Stata.

To create a new **Stata** icon you just copy an existing **Stata** icon:

1. Click on the existing **Stata** icon.

2. Pull down **File** and select **Copy**.

3. Set the **To Group**. You can use the group containing the original **Stata** icon if you wish.

4. Click **OK**.

You now have two icons. You can change the memory or start-up directory of the new icon just as you would the old; see [GSW] **B.4 The Windows 3.1 Properties Sheet**.

B.9   Running simultaneous Stata sessions

Each time you click on the **Stata** icon, you invoke a new instance of Stata, so if you want to run multiple Stata sessions simultaneously, you may.

If you are going to run simultaneous Stata sessions, it is best if you have installed `share.exe` into your DOS operating system. `share.exe` provides file-sharing and locking capabilities and came with your operating system.

Notes

C Troubleshooting starting and stopping Stata

Contents:
C.1   If Stata does not start
C.2   verinst problems
C.3   Troubleshooting

C.1   If Stata does not start

You tried to start Stata and it refused; Stata or your operating system presented a message explaining that something is wrong. Here are the possibilities:

Cannot find license file
This message means just what it says; nothing is too seriously wrong, Stata simply could not find what it is looking for, probably because you did not complete the installation process or Stata is not installed where it should be.

Did you insert the codes printed on your paper license to unlock Stata? If not, go back and complete the installation; see Chapter 1.

Assuming you did unlock Stata, Stata is merely mislocated or the location has not been filled in.

Error opening or reading the file
Something is distinctly wrong and for purely technical reasons. Stata found the file it was looking for, but either the operating system refused to let Stata open it or there was an I/O error. About the only way this could happen would be a hard-disk error. Stata technical support will be able to help you diagnose the problem; see [U] **2.8 Technical support**.

License not applicable
Stata has determined that you have a valid Stata license, but it is not applicable to the version of Stata which you are trying to run. You would get this message if, for example, you tried to run Stata for Windows using a Stata for Macintosh license.

The most common reason for this message is that you have a license for Small Stata, but you are trying to run Intercooled Stata. If this is the case, reinstall Stata, making sure to choose the appropriate version.

Other messages
The other messages indicate that Stata thinks you are attempting to do something you are not licensed to do. Most commonly, you are attempting to run Stata over a network when you do not have a network license, but there are a host of other alternatives. There are two possibilities: either you really are attempting to do something you are not licensed to do or Stata is wrong. In either case, you are going to have to call us. Your license can be upgraded or, if Stata is wrong, we can provide codes over the telephone to make Stata stop thinking you are violating the license. See [U] **2.8 Technical support**.

C.2   verinst problems

Once Stata is running, you can type `verinst` to check if it is correctly installed. If the installation is correct, you will see something like

```
. verinst
You are running Intercooled Stata 6.0 for Windows.
Stata is correctly installed.
You can type exit to exit Stata.
```

If, however, there is a problem, `verinst` will report it.

In most cases, `verinst` itself tells you what is wrong and how to fix it. There is one exception:

```
. verinst
unrecognized command
r(199);
```

This most likely means that Stata is not correctly installed. If you need to install Stata in a nonstandard way, you can learn exactly how Stata works by reading [GSW] **A. Starting and stopping Stata for Windows 98/95/NT** or [GSW] **B. Starting and stopping Stata for Windows 3.1**. Otherwise, reinstall Stata following the instructions in Chapter 1.

C.3   Troubleshooting

Crashes are called Application Faults in Windows. The dreaded Application Fault is typically not the application's fault. Most commonly, it is caused by configuration problems, bugs in device drivers, memory conflicts, and even hardware problems.

If you experience an Application Fault, first look at the Frequently Asked Questions (FAQ) for Windows in the user support section of the Stata web site *http://www.stata.com*. You may find the answer to the problem there. If not, we can help, but you must give us as much information as possible.

Reboot your computer, restart Stata, and try to reproduce the fault, writing down everything you do before the fault occurs. We will want that information along with the contents of your `CONFIG.SYS`, `AUTOEXEC.BAT`, `SYSTEM.INI`, and `WIN.INI` files. If you cannot email them to us, at least print them so you can look at them when you call us.

If Stata used to work on your computer but suddenly stopped working, try to remember any hardware or software you have recently installed.

In addition, give us as much information about your computer as possible. Is it a Pentium? A 486? Are you running Windows 3.1, 95, 98, or NT? What brand is your computer? What kind of mouse do you have? What kind of video card?

Finally, we need your Stata serial number and the date your version of Stata was "born". Include it if you email, know it if you call. You can obtain it by typing `about` in Stata's Command window. `about` lets us know everything about your copy of Stata, including the version and the date it was produced.

D Setting the size of memory

Contents:

D.1 Memory size considerations

Stata works with a copy of the data that it loads into memory.

By default, Small Stata allocates about 400K to Stata's data areas and you cannot change it.

By default, Intercooled Stata allocates 1 megabyte to Stata's data areas and you can change it.

You can even change the allocation to be larger than the physical amount of memory on your computer because Windows provides virtual memory.

Virtual memory is slow but adequate in rare cases when you have a dataset that is too large to load into real memory. If you use large datasets frequently, we recommend that you add more memory to your computer. See [GSW] **D.4 Virtual memory and speed considerations**.

One way to change the allocation is when you start Stata. This was discussed in

> Windows 98/95/NT   [GSW] **A.6 Specifying the amount of memory allocated**
> Windows 3.1   [GSW] **B.6 Specifying the amount of memory allocated**

In addition, if you use Stata for Windows 98/95/NT, you can change the total amount of memory allocated while Stata is running. That is the topic of this chapter.

Understand that it does not much matter which method you use. Being able to change the total on the fly is convenient, but even if you cannot do this, it just means you specify it ahead of time and if later you need more, you must exit Stata and reinvoke it with the larger total.

D.2 Setting the size on the fly: Windows 98/95/NT

Assume that you have changed nothing about how Stata starts so you get the default 1 megabyte of memory allocated to Stata's data areas. You are working with a large dataset and now wish to increase it to 32 megabytes. You can type

```
. set memory 32000
(32000k)
```

and, if your operating system can provide the memory to Stata, Stata will work with the new total. You type `set memory 32000` because you specify the total in K: 32,000K is roughly 32 megabytes. Later in the session, if you want to release that memory and work with only 2 megabytes, you could type

```
. set memory 2000
(2000k)
```

189

There is only one restriction on the set memory command: whenever you change the total, there cannot be any data already in memory. If you have a dataset in memory, you save it, clear memory, reset the total, and then use it again. We are getting ahead of ourselves, but you might type

```
. save mydata, replace
file mydata.dta saved
. drop _all
. set memory 32000
(32000k)
. use mydata
```

When you request the new allocation, your operating system might refuse to provide it:

```
. set memory 128000
op. sys. refuses to provide memory
r(909);
```

If that happens, you are going to have to take the matter up with your operating system. In the above example, Stata asked for 128 megabytes and the operating system said no.

D.3   The memory command

memory helps you figure out whether you have sufficient memory to do something.

```
. memory
```

Total memory	1,048,576 bytes	100.00%
overhead (pointers)	114,136	10.88%
data	913,088	87.08%
	------------	
data + overhead	1,027,224	97.96%
programs, saved results, etc.	1,552	0.15%

Total	1,028,776	98.11%
Free	19,800	1.89%

19,800 bytes free is not much. You might increase the amount of memory allocated to Stata's data areas by specifying set memory 2000. Stata for Windows 3.1 users would need to exit Stata and change the k option in the properties for **Stata**'s icon.

```
. save nlswork
. set memory 2000
(2000k)
. use nlswork
(NLS Women 14-26 in 1968)
. memory
```

Total memory	2,048,000 bytes	100.00%
overhead (pointers)	114,136	5.57%
data	913,088	44.58%
	------------	
data + overhead	1,027,224	50.16%
programs, saved results, etc.	1,552	0.08%

Total	1,028,776	50.23%
Free	1,019,224	49.77%

Over 1 megabyte free; that's better. See [R] **memory** for more information.

D.4 Virtual memory and speed considerations

When you open (`use`) a dataset in Stata for Windows, Stata loads the entire data into memory. If you have Intercooled Stata, you can change the amount of memory Stata allocates; see the beginning of this chapter.

Stata can use virtual memory under Windows. This allows you to use more memory than exists on your system. Windows substitutes hard disk space in place of memory. Stata does not know when Windows does this—Stata merely asks Windows for some amount of memory, and Windows either provides it or refuses. It is entirely up to Windows whether the memory provided is real or virtual. Windows will use virtual memory when the total memory demand across all active tasks exceeds the physical total on your computer. The more memory you allocate to Stata, the more likely Windows is to use virtual memory.

You want to avoid making Windows use virtual memory except when you really need it. Virtual memory is fine for rare situations where you must work with a dataset that will not fit into real memory (RAM), but it is not a viable alternative for daily use.

What many users do not realize is that merely allocating a lot of memory to Stata may cause Windows to start using virtual memory (paging) even if Stata is not using all that memory. From time to time we hear from users who say something like "My computer has 32 megabytes of RAM and I'm only working with a 2 megabyte dataset, but Stata is going really slow and my hard disk is spinning a lot." Invariably we discover that the user has allocated 32 MB of memory to Stata—many times more than what is necessary.

If you have a 32 MB computer and are running Windows 98 or 95, the maximum amount of memory that you should allocate is roughly 22 MB. Allocate more than that and Windows will start paging. The relationship between the amount of RAM and when paging begins is not linear. On a 16 MB computer, allocate more than around 9 or 10 MB and Windows will start paging. This is true even if you use small datasets. However, the probability that Windows pages on a given task is reduced with smaller datasets. As Windows NT users are probably already aware, Windows NT consumes even more resources than Windows 98 or 95, so you will experience paging even sooner under Windows NT.

For Windows 3.1, it is simpler. The maximum amount of memory you should allocate is roughly 4 MB less than the amount of RAM. For example, on a 16 MB computer, allocate more than about 12 MB and Windows 3.1 will start paging.

The above assumes that Stata is the only application running. If you want to run other applications simultaneously, you should reduce the amount of memory allocated to Stata.

When you use more memory than is physically available on your computer, Stata slows down. If you are using only a little more memory than is on your computer, performance is probably not too bad. On the other hand, when you are using a lot more memory than is on your computer, performance will be noticeably affected. In these cases, we recommend that you

```
. set virtual on
```

Virtual memory systems exploit locality of reference which means that keeping objects closer together allows virtual memory systems to run faster. `set virtual` controls whether Stata should perform extra work to arrange its memory to keep objects close together. By default, `virtual` is set `off`. `set virtual` can only be specified if you are using Intercooled Stata for Windows.

In general, you want to leave `set virtual` set to the default of `off`. Stata will run faster.

When you `set virtual on`, you are asking Stata to arrange its memory so that objects are kept closer together. This requires Stata doing a substantial amount of work. We recommend setting virtual on only when the amount of memory in use drastically exceeds what is physically available. In these cases, setting virtual on will help, but keep in mind that performance will still be slow. See [U] **7.4 Virtual memory and speed considerations**.

In summary, you may have to experiment to find the maximum amount of memory you can allocate to Stata and not have Windows start paging. This is the most you will want to allocate to Stata to do your day-to-day work. Only set the amount of memory close to or above the physical amount of RAM on your computer in emergencies. For example, you may want to read in a very large dataset just to split it into pieces.

If you find that you are experiencing many "rare situations" where you must work with a dataset that will not fit in real RAM, we recommend that you purchase more memory. Memory is inexpensive and you can generally find it cheaper from sources on the web than in catalogs. Start by searching for "`memory chips`" using a search engine.

E More on Stata for Windows

Contents:

E.1   Using Stata datasets and graphs created on other platforms
E.2   Importing a Stata graph into another document
E.3   Saving the contents of the Review window as a do-file
E.4   Making Windows 98/95/NT show file extensions

E.1   Using Stata datasets and graphs created on other platforms

Stata will open any Stata `.dta` dataset or `.gph` graph file regardless of the platform on which it was created, even if it was a Macintosh or Unix system. In addition, Stata for Macintosh and Stata for Unix users can use any files you create. If you transfer a Stata file using file transfer protocol (FTP), just remember to transfer using binary mode rather than ASCII.

If you need to send a dataset to a colleague who is using Stata 5.0, first, recommend that they upgrade to the latest version of Stata. Then, save your dataset with the `old` option so they will be able to read it even if they do not upgrade. For example,

```
save "c:\my data\flood2.dta", old
```

E.2   Importing a Stata graph into another document

The easy way to import a Stata graph into another application is via the Windows clipboard.

Create your graph. To copy it to the clipboard, pull down **Edit–Copy Graph**. By default, Stata will copy the graph as a Windows Metafile (WMF). This is probably what you want; it ensures that the receiving application obtains it in the highest resolution possible. If you want to copy it as a bitmap, however, you can pull down **Edit–Graph Copy Options** and then set the style to **Windows Bitmap**.

In general, you will want to copy graphs as metafiles. A bitmap is simply a picture, but a metafile contains the commands necessary to redraw the graph. That is, a metafile is a collection of lines, points, text, and color information. Metafiles therefore can be edited in a structured drawing program.

You will also probably want Stata to use Windows TrueType fonts, which is what Stata does by default. You can switch to Stata's built-in font by unchecking **Use Windows fonts** under **Prefs–Graph preferences**, but we recommend that you leave this preference checked. See [G] **printing Windows** for a complete discussion of the use of Windows fonts in Stata graphs.

After you have copied the graph to the clipboard, switch to the application into which you wish to import the graph and paste it. In most applications, this is accomplished by pulling down **Edit–Paste**. Consult the documentation for your particular application for more details.

Stata also gives you the ability to create a file containing the metafile. You could later import this file into another application. To save a graph as a metafile, pull down **File–Save Graph**. In the **Save** dialog box, choose **Windows Metafile** as the file type.

The `gphprint` command can also save graphs as metafiles. `gphprint` can also save encapsulated PostScript (EPS) files if you set your default printer driver to create EPS. EPS files can also be imported into desktop publishing applications. See [G] **printing Windows** for more information.

E.3   Saving the contents of the Review window as a do-file

To save the contents of the Review window as a do-file, select **Save Review Contents** from the Review window's system menu (the little box at the left of the Review window's title bar). Stata will save the commands in the file you specify in the **Save** dialog. You can edit the do-file in Stata's do-file editor (see Chapter 14 earlier in this book) if you are using Stata for Windows 98/95/NT or in a text editor such as Notepad if you wish. If you use a word processor to edit the file, make sure to save the file as text. Also, be aware that some editors such as Notepad will always try to append a `.txt` extension to the filename. Even though you may have typed `myfile.do` in the **Save as** dialog box, Notepad will save `myfile.do.txt`. To force such an editor to use the extension you want, enclose the filename in double-quotes; type `"myfile.do"` rather than `myfile.do`.

For more information on do-files, see [U] **19 Do-files**.

E.4   Making Windows 98/95/NT show file extensions

When you look at files using Windows Explorer, the file extensions may be hidden from you. This makes it difficult to rename files. For example, if you wish to rename `myfile.txt` to `myfile.do`, but extensions are hidden, you will only see `myfile` as the filename. When you click on `myfile` and type `myfile.do`, Windows will not change the extension on the file. Rather, since the `.txt` extension is hidden, you will now have a file named `myfile.do.txt`, with `myfile.do` showing and the `.txt` hidden.

You can turn off this behavior:

1. Double-click on the **My Computer** icon.

2. Pull down **View** and choose **Options...**.

3. Click on the **View** tab in the **Options** dialog box.

4. Make sure that **Hide MS-DOS file extension for file types that are registered** is not checked.

5. Click on **OK**.

Now, you can see the full names of files in Windows Explorer. When you click on `myfile.txt` and change its name to `myfile.do`, you will not have to worry about a hidden extension remaining on the file.